Teaching
Science
Today

2nd Edition

Author
Kathleen Kopp, M.S.Ed.

Foreword
Alan McCormack, Ph.D.

SHELL EDUCATION

Image Credits

p.86, 109, Kathleen Kopp; all other images Shutterstock

Shell Education

5301 Oceanus Drive
Huntington Beach, CA 92649-1030
http://www.shelleducation.com

ISBN 978-1-4258-1209-6

© 2015 Shell Educational Publishing, Inc.

Teaching
Science
Today
2nd Edition

Table of Contents

4

Foreword

Our ability to compete as a nation depends on knowledge, ingenuity, and a growing menu of intellectual skills. Scientific literacy is now paramount in a complex, competitive world and this demands excellence in our nation's science education. We desperately need fresh approaches to science education that fuse the Next Generation Science Standards, Common Core Standards, and major efforts to develop higher-order thinking skills in all of our students. Our 21st Century society and its supporting economy need to be refocused on up-to-the-minute scientific knowledge, innovation, and creativity.

It is widely known that there is an achievement gap between U.S. students and their international counterparts. American students fare poorly on international assessments such as the Programme for International Student Assessment (PISA). These assessments measure more than just mastery of memorized information—they assess higher-order thinking and problem-solving, which has been a stumbling block for our U.S. students. There is an important point here: Countries that do well on the PISA tests tend to demonstrate greater increases in Gross Domestic Product (GDP) than counties that perform less well.

In recent decades, our industrial economy has shifted to a service economy driven by need to use information and innovative skills rather than carry out repetitive manufacturing tasks. We now have a new, globally-connected economy, and companies have changed their modes of operation. Technology displaces workers with low-level skills who perform routine tasks. On the other hand, the new technology rewards workers with higher-level skills, enabling them to be more creative and productive. Workers need be able to respond flexibly to complex problems, work amicably in cooperative teams, deal with complex data, communicate effectively, and produce new ideas.

In *Teaching Science Today*, Kathleen Kopp elaborates the current need for retooling U.S. science education, noting the societal and economic changes mentioned above. She proposes that teachers of science at all grade levels incorporate the exciting new foci now revitalizing teaching and learning of science. She builds a strong case for the need for scientific literacy, aiming for commonplace skills among citizens to understand and evaluate science information as presented by the media, and be able to apply knowledge of human use of soils, water, air, and energy in sustainable ways. A scientifically literate society requires all citizens to be able to process scientific information and use it to make sound decisions regarding everyday personal life and world-changing decisions through thoughtful voting.

The author shows the key functioning of inquiry as the lifeblood of both professional science and as a primary mode for student learning of science. She guides teachers in the use of the inquiry-oriented 5E (Engage/Explore/Explain/Extend/Evaluate) Instructional Model, which has been thoroughly tested by both classroom practitioners and researchers and established as eminently effective. This approach is melded with both the New Generation Science Standards and the Common Core Standards showing how students can be inspired to be scientific problem-solvers, budding engineering designers, and also learn to better deal with information on the printed page, communicate better through writing and speaking and understand the uses or mathematics in dealing with personally-collected experimental data.

I think you will find this book to be a wonderful overview of the best current trends in science education and as a source for inspiration for making your classroom a place for joyful learning.

—Alan McCormack, Ph.D.
Professor Emeritus of Science Education
San Diego State University
Past President NSTA

Acknowledgments

I must thank my family for awarding me the time and space to put this book together for the benefit of science teachers everywhere. My two boys, Jeffrey and Christopher, are especially helpful and understanding when I drag out equipment to conduct mini-labs, clog our refrigerator with investigative samples, and go on hikes and road trips to gather scientific evidence and materials (e.g., rocks), which I use to develop engaging and meaningful lessons. These lessons I now pass onto you, the reader, for your own classroom.

This book is made possible in part because of the many teachers who are committed to providing engaging, meaningful science activities to their students in spite of time and financial shortfalls in today's schools. I enjoy teaching and learning right along with my peers. I appreciate opportunities to collaborate in my school and in my district, and on the web. I appreciate my school's support of science through the benefit of a full-time science resource teacher and the encouragement of field trips and guest speakers to spark students' interest in science. I also have had many, many opportunities to continue my own professional development in the area of science and the teaching of science. These new skills and strategies weave easily into other areas of the curriculum. With today's emphasis on reading, writing, and math, I gratefully present ideas for science teachers to support these essential areas of the curriculum. Together, teachers of all subjects can continue to move students through the 21st century and help students regain their lost ground internationally in science.

I also appreciate my community's commitment to science and scientific learning. First, thank you, Citrus County Schools, for supporting the district's Marine Science Station and Science Fair competition. Thank you, Florida, for protecting ecologically sensitive lands and supporting our State Parks system. Thank you, Southwest Florida Water Management District for sponsoring annual science-related grants. Finally, thank you to our area businesses and resident experts who eagerly visit schools and book field trips in the name of science learning. Your care and concern to build a future of scientists surely help teachers to help students see value and gain interest in science-related topics.

—Kathleen Kopp

Introduction

Science education in the United States today needs your help. Our students are graduating from high school without understanding much about the world, and they are lagging far behind graduates in other countries in problem-solving, mathematics, and reading skills (Layton 2013). Fewer and fewer students in the United States are choosing careers in science and engineering (Hossain and Robinson 2012). The dire lack of women and minorities in science research and applications fields is concerning (Pollack 2013).

Yet, nearly all aspects of daily living today, and in the foreseeable future, require an understanding of basic and advanced science and technology concepts and skills. *What is global warming all about? What are stem cells, and why do scientists want to study them? How do I know this medicine is safe and effective? Should I vaccinate my child? What are alternative energy sources? Will eating a genetically modified food change my own DNA or cause me harm?* These are everyday questions asked in today's society. Who knows what questions will become imperative tomorrow, and how will we be able to answer them?

The importance of science in daily life was highlighted when a major tsunami hit several countries surrounding the Indian Ocean on December 26, 2004, resulting in nearly 230,000 deaths. One 10-year-old girl is credited with saving the lives of her family and approximately 100 tourists staying at the same hotel. When she saw the water rush out, and large waves beginning to violently return, she warned her mother that these were signs of a tsunami and that they should get off the beach immediately. Her mother quickly notified the hotel management, who evacuated the hotel and beachfront. Two weeks earlier, the girl had completed a school project on tsunamis (The Telegraph 2005). Her teacher could not have imagined this being a result of his lessons on a relatively rare phenomenon, but it illustrates the importance of teaching students facts and concepts about the world in which they live.

You will never know if one lesson, one demonstration, or one project you conduct with students will start them on the path to becoming research scientists, engineers, doctors, astronauts, storm chasers, or members of any other science-related career. However, many professionals in science can point to a single moment in time, or a single encounter with one person, that marked the beginning of a lifelong love of science. It doesn't have to be you, but it *could* be you.

To remain globally competitive, the United States needs competent, qualified scientists and engineers. In an article posted by the Center for American Progress, Donna Cooper, Adam Hersh, and Ann O'Leary acknowledge the growing share of economic output by China and India between 1980 and 2011. They explain that the commitment of these countries to education that focuses on science, math, engineering, and technology (commonly referred to as STEM) courses is "an integral part of a national economic strategy" (Cooper, Hersh, and O'Leary 2012, under "The Competition that Really Matters"). However, students in the United States continue to lag behind their global peers in the areas of science and math. The Huffington Post (2012), using evidence from a report published by Harvard University's Program on Education Policy and Governance, states that students in other countries are making academic gains two and three times the rate of students in the United States. Students in the United States ranked 29th out of 34 countries in math and science. Tomorrow's economic leaders all begin as children in classrooms. It could be *your* enthusiasm, *your* ability to provide interesting learning activities, and *your* encouragement to develop critical-thinking and problem-solving skills that will make all the difference. Maybe someday you'll meet an adult who greets you with, "Remember me? I was in your class. I became a (e.g., doctor, veterinarian, nutritionist, meteorologist, researcher, teacher) because of what I did in your classroom." Teachers can ask for no more greater reward than this.

You may be a beginning teacher, a veteran teacher, or a teacher who wants to personally enhance his or her instructional approach to teaching science concepts, skills, and processes. Regardless of who you are or where you are along your career path, by the time you finish this book, you will surely have many meaningful and engaging ideas up your sleeve to adequately and thoroughly provide science lessons to intrigue, motivate, and, teach what our students need today to be productive citizens of tomorrow.

Chapter 1

Science Education: Past, Present, and Future

 "Men love to wonder, and that is the seed of science."

—Ralph Waldo Emerson

Like all topics in education, science education has experienced a swing from one extreme to the other—from exclusively text-based instruction to exclusively hands-on activity. Is any one approach better than another? What about safety issues and concerns about students' maturity levels? How do these approaches acknowledge and accommodate students with special needs or those who are English language learners? Before seeing where we are today with regard to science education, let us take a short trip through the past.

Science Then

The first mention of United States training in hands-on experiences and physical examples for teaching students (known at the time as the "object method") dates back to 1861 (Sykut 2005). The Oswego Primary Training School (at the time, an industrial arts school) was founded in this year, and these instructional methods were led by Margaret E. M. Jones from London (LaValley). With the advent of the industrial revolution, teachers readily adjusted their instructional practices to meet the needs of high-demand skills and labor. These practices were better conducted through interactive, hands-on experiences and opportunities rather than through reading about how things work in manuals.

Since the mid- to late-nineteenth century, philosophers and science advocates have been imposing their thoughts and ideas regarding science education on the American education system. Biologist and author Thomas H. Huxley and philosopher Herbert Spencer, as well as Horace Mann and John Dewey, all have had influences on education in general, including science education.

Education has been influenced by worldwide events in addition to theorists and philosophers. Fast forward to the twentieth century. The launch of *Sputnik*, the world's first artificial satellite by the Soviet Union on October 4, 1957, challenged the standing of the United States as a global leader in science and technology. That single successful launch led to two major changes in how science and technology were emphasized at the national level. The first change in the United States was the establishment of the National Aeronautics and Space Administration (NASA) in 1958. The second change, in the same year, was the passage of the National Defense Education Act (NDEA). This federal legislation was designed to increase the academic competitiveness of students in the United States in mathematics, science, and foreign languages (Siegel 1963).

As we moved into the twenty-first century, societal topics, such as global warming and deforestation, and technological advances, such as electric cars and mobile devices, surely impacted what the science teachers chose to teach. However, it wasn't until recently that science education standards have been an area of national importance. For example, the Next Generation Science Standards (NGSS) are focused on "preparing students for college and careers (2013a, 2). These standards are "goals, that reflect what a student should know and be able to do (2013a, 2)."

Along with new theories and philosophies came changes in education: what students should know, how teachers should teach, and how students learn best. Fortunately for us today, teachers and students have been exposed to a wide variety of instructional approaches and learning models through the years. Theories and philosophies have morphed into research and best-practice models. We know what to teach, we know how best to teach it, and we know how to support student learning to maximize achievement. Of course, teachers are learning more each day, but we have a good foundation upon which to build.

Today's Reality

When measuring the science skills of fourth-grade students in the United States and their peers in other countries, the *Trends in International Mathematics and Science Study* (TIMSS) revealed that students in the United States were outperformed by students in only five other countries (Singapore, Chinese Taipei, Japan, Hong Kong, and England). By the time these students reached 8th grade, their ranking dropped significantly. By graduation, these same students ranked near the bottom of all students in math and science skills (TIMSS 2011). It appears that students show an interest and propensity for learning science at a young age, but this interest and ability precipitously decline as they move through the United States educational system.

A report titled *Rising Above the Gathering Storm* was released by a committee of experts from K–12 education, higher education, research and development companies, top-ranking scientists, and members of the National Academy of Sciences (2007). This report states, "Having reviewed trends in the United States and abroad, the committee is deeply concerned that the scientific and technological building blocks critical to our economic leadership are eroding at a time when many other nations are gathering strength. We fear the abruptness with which a lead in science and technology can be lost—and the difficulty of recovering a lead once lost, if indeed it can be regained at all" (National Academy of Sciences 2007, 3).

The Program for International Student Assessment (PISA) is an international assessment coordinated by the Organization for Economic Cooperation and Development (OECD). It measures 15-year-old students from around the world in terms of reading, math, and science literacy skills. PISA collects, organizes, analyzes, and publishes the results in all these areas. PISA also includes measures of general cross-curricular competencies, such as problem solving. In 2003, PISA surveyed student knowledge and problem-solving skills of 250,000 students in 40 countries. It assigned students achievement levels based on raw scores, with levels defined by the following descriptions:

- **Level 3:** Reflective, communicative problem solvers
- **Level 2:** Reasoning, decision-making problem solvers
- **Level 1:** Basic problem solvers
- **Below Level 1:** Weak or emergent problem solvers

In the United States, 58 percent of 15-year-olds who took the test scored at or below level 1—weak, emergent, or basic problem solvers. Based on the realities of today's global workplace, economists believe that these students are the ones who will not be able to compete in an increasingly technological, highly skilled workforce. Perhaps of even more concern is how students in the United States compare to those in other countries. Of the 40 countries assessed in the PISA study, the United States, once a leader in education, ranked 29th in the percentage of students achieving levels 2 or 3 in problem-solving abilities. This suggests that current methods of education do not allow United States students to effectively learn essential problem-solving skills. The educational system we have been using since prior to World War II is not meeting the needs of students preparing to be college and career ready for the 21st century. There are many explanations for this mismatch of education and workforce skills, but the fact remains that students in the United States must gain the skills to be more competitive in tomorrow's workforce, especially with regard to problem solving. The United States answered the call by creating the Next Generation Science Standards (NGSS 2013g). These standards will help students be better prepared for the future.

Today, the United States faces a critical need for its students to begin meeting the higher levels of achievement of their global peers in science, mathematics, and problem-solving skills. Because students in the United States are underperforming in these areas, teachers can sense there is an immediate need for significant improvement in science education. The basic skills necessary for all students entering today's workforce are at the very heart of scientific investigation. One of the most important ways students can learn about science is through collaborating to tackle real-world problems. By doing so, students can gain the skills they need to analyze and apply information across the curriculum. This type of inquiry-based learning is the foundation of current science education reform.

What's New in Science Education?

The National Research Council (NRC) developed its landmark publication, the *National Science Education Standards*, in 1996. On April 10, 2012, a 26-state consortium released the *Next Generation Science Standards* (NGSS). These latest standards are based on a report generated by the NRC: *A Framework for K–12 Science Education: Practices, Crosscutting Concepts, and*

Core Ideas. The NGSS refer to the Practices, Crosscutting Concepts, and Core Ideas as the "three dimensions."

Practices

The Science and Engineering Practices encourage teachers to think of scientific investigations as a combination of skills (abilities to perform tasks) and knowledge (information). These include the following:

- Asking Questions and Defining Problems
- Planning and Carrying Out Investigations
- Analyzing and Interpreting Data
- Developing and Using Models
- Constructing Explanations and Designing Solutions
- Engaging in Argument from Evidence
- Using Mathematical and Computational Thinking
- Obtaining, Evaluating, and Communicating Information

Crosscutting Concepts

The Crosscutting Concepts link all the domains in the core ideas. These concepts include the following:

- Patterns
- Cause-and-Effect Relationships
- Scale, Proportion, and Quantity
- Systems and System Models
- Energy and Matter: Flows, Systems, and Conservation
- Structure and Function
- Stability and Change

Core Ideas

There is no way any one teacher or series of teachers can possibly teach students every piece of scientific information. According to the NGSS Executive Summary, the goal of science education is to "prepare students with sufficient core knowledge so that they can later acquire additional information on their own" (2013a, 2). The Disciplinary Core Ideas relate to the following:

- Physical Science
- Life Science
- Earth and Space Science
- Engineering, Technology, and the Application of Science

According to the National Science Education Standards (National Research Council 1996), regardless of grade level, background, or intellectual ability, all students are expected to emerge from their thirteen years of schooling being able to "ask, find, or determine answers to questions derived from curiosity about everyday experiences... describe, explain, and predict natural phenomena... read with understanding articles about science in the popular press and to engage in social conversation about the validity of the conclusions... identify scientific issues underlying national and local decisions and express positions that are scientifically and technologically informed... evaluate the quality of scientific information on the basis of its source and the methods used to generate it... pose and evaluate arguments based on evidence and to apply conclusions from such arguments appropriately" (22).

In addition to the newly published National Science Education Standards, many states have also adopted the Common Core State Standards for English/Language Arts and Mathematics. The Common Core English/Language Arts standards include specific standards related to reading and writing in science (and other subject areas). Likewise, since scientific equipment typically requires students to measure, and investigations require simple to complex computations with whole numbers, fractions, and decimals (and all their conversions), the mathematics standards also play a part in the overall education of today's science students.

Looking Ahead

What do these standards mean for science teachers? It means that they need to prepare today's students for a much higher level of understanding of science and technology than expected in previous generations. Numerous research studies by the North Central Regional Educational Laboratory (NCREL) demonstrate that the skills today's students will need to succeed in the twenty-first century workforce are different from the skills taught in the past. NCREL states: "As society changes, the skills needed to negotiate the complexities of life also change. In the early 1900s, a person who had acquired simple reading, writing, and calculating skills was considered literate. 'Only in recent years has the public education system expected all students to build on those basics, developing a broader range of literacies' (Educational Testing Services 2002). To achieve success in the twenty-first century, students also need to attain proficiency in science, technology, and culture, as well as gain a thorough understanding of information in all its forms" (NCREL 2003, 15).

According to Stuart W. Elliott, director of the Board on Testing and Assessment at the National Research Council, "If you're a K–12 teacher, the workforce that you're influencing is one that will exist several decades into the future, not the one that exists now. You need to shift your focus into the future" (Cech 2007, under "Job Skills of the Future in Researchers"). The best way for today's teachers to accomplish this seemingly impossible mission is to maintain focus on the Next Generation Science Standards; provide opportunities for students to engage in open-ended critical analysis scientific investigations and STEM activities; and integrate technology as much as possible.

Conclusion

Science education, like other subject areas, has undergone some major shifts. Today's science teachers face the daunting task of preparing scientifically literate students in an ever-changing, technology-driven society. No one knows exactly what the future holds for students, but today's teachers can use the best research-based, educationally sound instructional practices to prepare them for an exciting future in a global society.

The standards have been reviewed in this chapter; subsequent chapters will address scientific inquiry. Technology suggestions are peppered throughout the instructional strategies presented. By the end of this book, all teachers, regardless of their comfort level, experience, or expertise with teaching science, should be able to better provide for today's students for tomorrow's needs.

Stop and Reflect

1. Reflect on your personal experiences in science class as a student. What do you remember doing or learning about? How does this compare to what you want students to remember about your class?

2. Rate your state/district/school from 1–5 (rating: 1, all text; rating: 5, exclusively hands-on), with regards to the spectrum of approaches to science curriculum. Why do you rank them this way?

3. How do you see the English/language arts and mathematics standards affecting science instruction? What are your initial thoughts regarding how they "fit" into science curriculum?

4. Why do you think society places emphasis on the importance of science in today's classrooms? How do you think this might change over the next 50 years?

Chapter 2

Becoming Scientifically Literate

 "Science is a way of thinking much more than it is a body of knowledge."

—Carl Sagan

Teachers oftentimes are concerned about the misconceptions their students bring to the science classroom. Likewise, adults have misconceptions about various processes and concepts. Not everyone can know everything there is to know about science concepts. Elementary school teachers are typically generalists. They must comprehend reading material, know the pedagogy behind math, and have at least a basic understanding of science, social studies, and health topics. However, some science ideas, such as knowing that a balloon filled with air weighs more than the same balloon without air, may be lacking from their general science knowledge. As teachers move into middle school and high school, their subject matter becomes much more specialized. Middle and high school science teachers may be savvy when it comes to understanding life science, but they may not fully understand advanced chemistry concepts. It stands to reason, if teachers are not comfortable with the actual content they must teach, they may shy away from it or avoid it altogether. So how do teachers of the youngest to oldest students become scientifically proficient to be the best teachers they can be?

Common Misconceptions

Teachers who really know science may be surprised to discover that teachers in their buildings may have misconceptions about science topics that they would consider "elementary." The Harvard-Smithsonian Center for Astrophysics (1987) produced a video titled, "A Private Universe." In it, they show Harvard alumni and new graduates struggling to explain what causes the seasons. The respondents overwhelmingly deduce that Earth is closer to the sun in summer, thus causing the daily temperatures to rise. In fact, the people on Earth experience seasons due to the fact that it tilts on its axis at different times of the year. If Harvard graduates fail to understand even basic science concepts, we can only assume that the general populous, including teachers, may lack these understandings, as well.

What is important for teachers to realize is the need for learning or relearning of the basic science concepts that they teach their students. Teachers may be tempted to rely on their student textbook for information; however, this may lead to further misconceptions. When the American Association for the Advancement of Science completed an evaluation of nine common middle school science textbooks, not one was found to rate a "satisfactory" ranking in the quality of instructional support materials (Koppal 2002). Teachers who only use the information available in an elementary or middle-school science book are more likely to pass on to students inaccurate or incomplete science concepts. Likewise, Internet websites may also have inaccurate or incomplete science concepts. To ensure accuracy, websites should be vetted and affiliated with reputable and accredited universities and research institutions.

What Today's Science Teachers Should Know

Science teachers should have some measure of scientific literacy so that they may be confident in what they must teach their students.

According to the National Science Education Standards (National Research Council 1996, 22), a person is considered scientifically literate if he or she:

- can ask, find, or determine answers to questions derived from curiosity about everyday experiences;

- has the ability to describe, explain, and predict natural phenomena;

- can read with understanding articles about science in the popular press and to engage in social conversation about the validity of the conclusions;

- can identify scientific issues underlying national and local decisions and can express positions that are scientifically and technologically informed;

- can evaluate the quality of scientific information on the basis of its source and the methods used to generate it; and

- can pose and evaluate arguments based on evidence and apply conclusions from such arguments appropriately.

The first step to learning and re-learning is to ask, "What content needs to be taught that I am unsure of?" Teachers should begin with the science standards they need to teach. These may be the Next Generation Science Standards or State Standards. The administrator or lead curriculum teacher should be able to provide this information to teachers.

Curriculum aside, teachers should also acquaint themselves with the technology available to them and their students. A school may have digital scales, digital microscopes, or graphing calculators with temperature probes to take digital temperature readings. Once the technology tools are known, teachers may then need to learn how to interface the technology with a projection system to conduct whole-class instruction with these tools. A media, technology, and/or curriculum leader is a good resource to utilize when setting up these devices. To minimize downtime in the classroom, teachers should practice using new technology. This will help to eliminate lack of student engagement. In order for the technology to be effective in the classroom, the teacher or student must use it effectively.

To learn about next generation standards for science, reading/language arts, and math visit these websites:

- Next Generation Standards Science Standards: http://www.nextgenscience.org/
- Common Core State Standards: http://www.corestandards.org/

Science teachers should also be proficient with essential skills such as observing, measuring, computing (and other math skills), using charts and tables to organize data, classifying, making inferences, predicting, solving problems, and summarizing. All of these skills come into play when teaching students about scientific knowledge and application, but they probably don't all come into play at the same time. For example, a teacher may need to explain the angle of incidence when explaining why we have seasons. As the angle of incidence changes, students may record this in a table, then use this data to create a chart. Eventually, the teacher may expect students to demonstrate their understanding through a cartoon with speech bubbles explaining why seasons occur. In order for students to attend to this type of task successfully, the teacher will likely want to provide a completed summary for students to use as a model.

Along with the idea of knowing scientific skills, teachers may need to be proficient with certain equipment and tools, depending on the grade level(s) they teach. These can include: thermometers, microscopes, stopwatches, compasses, balance scales (single, double, and triple beam), spring scales, beakers and graduated cylinders, compasses, barometers, microscopes, petri dishes, test tubes, stop watches, Bunsen burners, hot plates, flasks, meter sticks, centrifuges, and telescopes. Some investigations include the use of various chemicals such as iodine and ammonium. Likely each school district has policies and procedures for using open flames, heating elements, live (and deceased) animals, and what can be harmful chemicals. Teachers should familiarize themselves with a copy of these policies so that current and future scientific investigations may continue. If teachers come across a motivating and meaningful investigation using equipment, supplies, or materials with which they are unfamiliar, they should conduct the investigation first to be sure that they may adequately and smoothly guide students through the activity during class time.

Finally, teachers should be proficient with skills such as observing, predicting, and inferring. To be successful with this, teachers should research an activity they wish to conduct with students. Likely other

> For more information about using chemical safety, see Chapter 3.

teachers have posted blogs, resources, or how-to video demonstrations that will assist with the vision of the investigation. This can help teachers feel comfortable with the expected outcomes, and help them gain confidence in the completion of the activity in their own classrooms. For example, if a teacher wants to conduct a saltwater density activity with students, he or she may read and download any number of lab directions, and/or watch any number of how-to videos online. Having read about and seen these experiments occur, a teacher who has yet to do this with a class may walk into the classroom on lab day feeling more prepared, more confident, and more relaxed than a teacher who tries the same activity with no prior experience. Included with this idea of knowing and being able to apply process skills is the idea of conducting a full scientific experiment. Many "labs" claim to be experiments. But, in order for an activity to truly be an experiment, it must begin with a testable question, provide for the collection of data or evidence, and conclude with a summary using the data or evidence to respond to the initial question. Most true experiments also include a control and variables. Depending on the subject matter and grade level of the students, teachers should know the difference between an inquiry-based experiment and a non-inquiry-based activity.

> These ideas of inquiry and non-inquiry are explored in more detail in Chapter 4.

Professional Development Opportunities for Teachers

Today's science teachers have a myriad of opportunities to learn about science content and best-practices. This section explores just a sampling of the opportunities that are available.

Summer Courses at Colleges and Universities

College courses can be expensive, but credits can be used to achieve or maintain certification and/or an advanced degree. Courses range in topics and duration, and some may be taken online. Teachers interested in this option can look for classes at a local college or university, or investigate online classes and programs from accredited universities across the United States.

Workshops and Research Opportunities for Teachers

Teachers may search for any one of several privately-, federally-, or state-funded programs. Typically, free and stipend-paying workshops have limited spaces that fill up fast. Some are local and short in duration while others require travel time and are longer in duration.

Private, Federal, and State Science Learning Opportunities

The National Science Teachers Association (NSTA) has several learning opportunities for teachers. These include, but are not limited to symposia, institutes, seminars, podcasts, online courses, books, and articles. These online learning experiences are designed for teachers to enhance their understanding of key content areas.

Project WET (Water Education for Teachers) is committed to teaching children, adults, and communities about water education. Likewise, Project WILD connects children to wildlife by providing education, resources, and learning opportunities for anyone who teaches children about conservation and environmental education.

The Helios Education Foundation has partnered with the University of South Florida to "build a teacher preparation pipeline that will place 80 science, technology, engineering and math (STEM) teachers in Hillsborough County public middle schools by 2017" (Helios Education Foundation 2013, under "Chapter 2"). Both Science Foundation Arizona (SFAz) and Helios Education Foundation have also selected several schools in this state for the Helios STEM School Pilot program.

In an online article, Jen Scott Curwood (2013) mentions a summer scientific field program for teachers interested in archaeology, history, and natural science at the University of Alabama's Museum of Natural History. This is an example of just one of many summer field work experiences offered by one university system. Likely, teachers will find many others if they search a university close to them.

The following is an example of a national park learning experience for teachers, listed in the same article mentioned above (Curwood 2013). Arcadia National Park and the College of the Atlantic (in Maine) both offer a summer camp with fieldwork, labs, and lectures related to all things science, including astronomy, botany, geology, and oceanography. Other county, state, and national parks likely offer other experiences to further teachers' continued learning and professional development.

Conferences and Conventions

For decades, conferences and conventions have been a traditional means of continuing education for teachers. They continue to provide teachers with a common interest to come together in a large forum to share, talk, collaborate, learn, and grow professionally. Depending on the organizers and presenters, they may or may not provide the information teachers are looking for. However, they can re-energize teachers in their respective fields. When attended with a colleague, they can be extremely beneficial due to the level of collaboration that tends to occur. The National Science Teachers Association (NSTA) hosts several conferences around the nation each year. Likewise, individual state science organizations (such as the Minnesota Science Teachers Association) offer annual conferences which may be closer to home.

School-Based Workshops

Several aforementioned science learning organizations may provide school- or district-based workshops exclusively for science teachers. The Activities in Math and Science (AIMS) Education Foundation is a non-profit organization committed to providing teachers with engaging hands-on activities and learning opportunities for both math and science. They conduct workshops related to enhancing teachers' content knowledge, skills, and classroom practices.

Reliable Web Resources

Teachers can access literally thousands of websites devoted to providing exemplary lesson plans, lab activities, and other resources for science concept development. It may be difficult for teachers to keep track of which links they prefer using in the classroom and finding activities and information appropriate for the grade level they teach. Sites that compile links to multiple resources are a good place to narrow down a search. Suggested reliable lesson resource sites include (see Appendix C for URLs):

- The National Oceanic and Atmospheric Association (NOAA)
- The National Aeronautic and Space Administration (NASA)
- The Jason Project
- National Geographic for Teachers

Peer Collaboration

Teachers should remember that they likely have expert, highly effective teachers who can teach any student Newton's three laws of motion using nothing more than three pennies and a rubber band. Teachers wanting to learn content or to improve their teaching practices might tap the resources they have right in their very own schools. These efforts begin with a simple question, "Hey, do you have any teaching tips or must-do activities related to _____?" Trusted colleagues can sometimes be the best resources available to other teachers. Teachers can simply begin a dialogue and keep the lines of communication open.

In addition to the personal connection, teachers can reach out to each other through blogs, digital posting sites, and other digital tools. Blogs include opinion articles, resources, and information for those who follow their blog. Some blogging sites allow for reflection and feedback where viewers can post their personal ideas in response to the blogger's post. Teachers interested in joining a blog could simply search something such as *science blogs for teachers*. New connectivity sites, such as Pinterest©, Edmodo and Edutopia™ pop up on a regular basis. These sites typically have loyal followers who join specific categories (in this case, science teachers). The people who join the groups post exemplary ideas, websites, and information to share for free with others.

Book Resources

There's nothing wrong with turning to good old-fashioned books to learn about new information or to refresh previously learned information. Teachers can choose to review literature on any science topic in pretty much any format that suits them best. Some science resources take the form of college-level texts (or something that looks and reads like a college-level text), listing topics and information to the minutest detail. Other book resources are more handbook–like in nature. They provide the basic information targeted to a specific grade level or set of grade levels, and include information appropriate for students (and their teachers) in a format that is easier to read than the more thorough texts. Yet, other resources are written just for teachers. Some books follow a format similar to this; they are intended to be used for professional development purposes. Other teacher books tend to include specific grade-level lesson plans and information along with student pages so that teachers may "pick up and go," and not have to necessarily put a whole lot of time into developing the lessons. Teachers may even gain insight to a few facts and information through student books, such as textbooks and leveled science readers or science-related magazines written with students in mind. There are countless resources covering a plethora of topics available to teachers.

Conclusion

It is helpful for teachers to be conscious of their own possible misconceptions related to scientific concepts. To help with this, many organizations provide relevant, interesting, and purposeful workshops, conferences, and professional development opportunities (real-life and virtual) to help teachers grow and learn, and to be the most effective science teachers they can be. Additionally, teachers have access to books, the Internet, and their own colleagues to continue their professional development with regard to science concepts and instructional strategies. All it takes is for a teacher to make that first step toward his or her continuing education.

Stop and Reflect

1. Do you consider yourself to be "scientifically literate?" Explain your thinking.

2. What misconceptions or lack of knowledge do you have about a particular science topic? What resources are available to you to learn more about this topic? Find out more so that you are fully comfortable with the subject matter.

3. What scientific equipment and/or materials are available to you at your school? Choose one lesser-known piece of equipment or material. How might you use it safely and effectively during an upcoming science topic?

4. What and where are your science standards? What standards are you most unfamiliar with; or which standards seem to be the most challenging to teach? Research resources available to you to bring this topic to life with students.

Chapter 3

Inside the Science Classroom

 "Safety is something that happens between your ears, not something you hold in your hands."

—Jeff Cooper

Organizing the classroom is one of the most important aspects of teaching. Without a plan for organization and safety within the classroom, learning most likely will not take place. This chapter will provide you with information about suggested equipment and important safety guidelines to follow within a science classroom.

Equipment and Supplies

Some science teachers are fortunate to have all or most of the equipment and supplies they need to conduct pretty much any investigation that matches their content standards. They may have adequate equipment and supplies to allow for each small group of students to conduct its own investigation. However, even if teachers do have the equipment and supplies on hand, they may not be readily available. This may require teachers to conduct more demonstration labs rather than allow for individual or group labs. This is not the ideal situation, but it does bring real learning into the classroom. Perhaps more concerning are situations where teachers and students lack adequate equipment and/or materials. Schools, science teams, and individual teachers have many options to acquire the needed equipment and/or supplies. This section explores these options.

Large discount retailers carry everyday kinds of materials at low costs, such as measuring cups, rubbing alcohol, string, batteries, flash lights, seeds, potting soil, foam or plastic cups, food coloring, salt, vinegar, baking soda, straws, balloons, and marbles. Teachers might also ask about school funds allocated specifically for science materials. Other options include requesting donations from retailers or parents. Sending home a list of everyday materials that are used in a science classroom could prove to be beneficial.

More advanced materials and supplies can be found online through a simple search. Common specialty items for science classrooms include beakers and graduated cylinders, digital scales, meter sticks, thermometers, stopwatches, hand lenses, safety goggles, and plastic lab aprons. Since most of these supplies come with larger price tags, teachers should ask about school funds, or try their Parent Teacher Association (PTA) or Parent Teacher Organization (PTO) to contribute funds to support instruction. Teachers who collaborate on purchases by pooling classroom funds will find that their budgets go farther than if each teacher makes separate purchases.

In addition to parents and PTAs or PTOs, retailers, local businesses or research facilities may be interested in donating used equipment or purchasing new equipment for schools.

Finally, science teachers can explore grant funding for needed equipment and materials. Many groups, both locally and nation-wide, offer grants for teachers for a variety of purposes. Some schools or school districts have grant coordinators who help facilitate this process with teachers. If a grant writer is not available, teachers can go online to find a long list of grant opportunities that are available. Organizations and foundations have money to spend to support education; science teachers should not ignore this potential support system. Teachers can simply search "science grants for schools" to find a list of grant options. The grant application may include a checklist of tasks, and they should include evaluation criteria upon which each grant will be judged. This is helpful information for teachers who may be writing a grant for the first time. Most grant providers look for clear, explicit, meaningful, and creative instructional ideas.

Suggested Resources for Science Equipment and Supplies

- school funds, classroom funds, or PTA/PTO
- donations from parents, retailers, businesses, or laboratories
- mobile labs
- online simulations
- grants

Science Safety in the Classroom

Teachers need to manage time, people, materials, and space in an inquiry-based classroom. Teachers should establish clear expectations for lab time. In addition, science teachers might also want students to take a lab group "safety oath," or sign a lab-group contract that states, among other things, that they will abide by lab safety procedures. Part of this process includes establishing consequences for non-compliance. Students should know what these consequences are, and teachers must follow through to ensure safety for everyone and everything in the classroom.

Chemical Safety

Some particularly strong acids "fume"—that is, they can turn from a liquid into a gas at room temperature, releasing poisonous, corrosive gases, even when stoppered. These gases can combine with metal, other chemicals, and even paper to create a fire hazard. Other chemicals, such as fabric softener, vinegar, and food coloring, are less hazardous. Regardless, every chemical used during science class deserves the proper care and consideration for the safety of the teacher, students, and the school. The Center for Disease Control and the American Chemical Society both have safety guides and procedure manuals available for free download on their websites. This literature helps teachers identify harmful and potentially harmful materials and guide their safe use in classrooms. Additionally, individual schools, school districts, or state education departments may have additional protocols in place regarding the use of chemicals in schools.

Teachers should be aware of allowable materials and abide by the safety regulations set forth by the governing body.

Chemicals come with cautionary labels that teachers should heed. Labels might include one of three levels of caution: *danger* (the strongest warning), *warning*, and *caution*. Additionally, labels may caution against the potential danger. These warnings may read: *flammable, harmful if inhaled, causes severe burns/irritation,* or *poison*. Additionally, labels will list precautionary measures to avoid these harmful risks. For example, a chemical with a label that reads, *keep away from open flame* is best stored away from electrical devices, especially Bunsen Burners. If for some reason something happens, labels will also likely have first aid information. Teachers should know the potential hazards and be prepared should warnings go unheeded by students.

When in doubt, teachers should use caution when introducing, storing, and using chemicals in the classroom. They should keep potentially harmful (and even non-potentially harmful) materials out of the reach of students until an investigation takes place. Teachers should store chemicals properly at all times. This requires an isolated room with a door (and preferably a lock) at the appropriate temperature. Some science materials are dangerous to use without proper protective gear (such as goggles and protective clothing) and adult supervision. Teachers could acquire a portable eyewash station if a sink is not readily handy. Teachers should strongly caution students to never touch, smell, and certainly not taste materials until told when and how to do so. Not every liquid can be safely poured down a sink, and not every solid can be tossed into a trash can. Therefore, teachers should properly discard used and unused chemicals after use.

Fire and Electrical Safety

Some interesting (not to mention visually exciting) science demonstrations involve the use of flames or electricity. Teachers should consider the fact that students may feel compelled to mimic, imitate, or otherwise try demonstrations on their own. Those that involve explosions, shooting flames, or electrical charges are especially appealing to recreate. Teachers should take care not to give out exact information on how to reproduce these events. They should also keep these important fire and electrical safety considerations in mind when using these types of energy. First of all, schools, school districts, or state departments of education may have restrictions involving the use of open flame or electrical appliances. Teachers should know these restrictions and take the proper precautions as dictated, such as moving to an approved area that allows electricity. When using open flames in a classroom, teachers should have the proper safety equipment readily available (e.g., fire extinguisher, fireproof screen, an emergency shutoff for gas valves if using gas, proper disposal containers for matches, safety goggles, hot pads for resting hot glassware upon, and fire-resistant table coverings). Open flames should not be used with younger students.

Older students should be taught to roll back long sleeves, tie back long hair, and wear proper safety equipment. Teachers must insist that students follow proper safety procedures, and they should implement the school's discipline policy if students fail to do so. Students who do not follow safety guidelines pose a threat to everyone around them, not to mention themselves.

> If a science classroom does not have its own fire extinguisher, the teacher should know where to find one quickly in the school building. It is extremely important to know where safety equipment is at all times!

Likewise, when teachers use electrically powered devices for science investigations, they must insist on rigorous safety guidelines. Even though students are accustomed to using electrical appliances, the classroom is a unique setting. Sometimes, electrical cords stretch across the floor where students can trip over them. In these instances, teachers should be mindful to limit movement around the room, and do what they can to minimize any potential hazards. Teachers can also use these moments to instruct students regarding fire and electrical safety. The Electrical Safety Foundation International has a fire and electrical safety "toolkit" (along with other resources such as videos, posters, and games) available for free download. This toolkit includes student activities and teacher-friendly lesson plans intended to maximize safety and minimize danger.

Animals in the Classroom

Live animals bring interest and excitement to a classroom. From worms and insects to caterpillars and tadpoles, students can learn valuable lessons regarding health and safety risks and care and compassion when teachers incorporate animals into their curriculum. Classrooms may have a class pet for students to help care for such as, frogs, fish, snakes, hamsters, mice, or guinea pigs. In upper-grade classes, students look forward to (perhaps with squeamish excitement) the prospect of dissecting animals to learn firsthand about life cycles and body functions. Although the integration of animals may not seem risky on the surface, teachers should understand that this unique process likely comes with expectations, rules, and restrictions.

As with many other science-education issues, the National Science Teachers Association (NSTA) has issued a position statement on the use of live animals in the science classroom: "NSTA supports the decision of science teachers and their school or school district to integrate live animals and dissection in the K–12 classroom. Student interaction with organisms is one of the most effective methods of achieving many of the goals outlined in the National Science Education Standards. To this end, NSTA encourages educators and school officials to make informed decisions about the integration of animals in the science curriculum. NSTA opposes regulations or legislation that would eliminate an educator's decision-making role regarding dissection or would deny students the opportunity to learn through actual animal dissection.

> "NSTA encourages districts to ensure that animals are properly cared for and treated humanely, responsibly, and ethically. Ultimately, decisions to incorporate organisms in the classroom should balance the ethical and responsible care of animals with their educational value."

—NSTA 2013, under "NSTA Position Statement"

Teachers wanting to include animals as part of their science studies should check with the principal or department head first to see what rules, restrictions, or protocols are in place. Different states may also have their own rules about animals in public schools. Even with restrictions, teachers may have options to bring in certain animals temporarily. Some animals, while typically banned from school, may be used in investigations without violating policies regarding animals in the classroom. In addition, the following important points for teachers to remember come from the NSTA position statement.

Teachers should:

- educate themselves about the safe and responsible use of animals in the classroom

- seek information from reputable sources and familiarize themselves with laws and regulations in their state

- become knowledgeable about the acquisition and care of animals appropriate to the species under study so that both students and the animals stay safe and healthy during all activities

- follow local, state, and national laws, policies, and regulations when live organisms, particularly native species, are included in the classroom

- integrate live animals into the science program based on sound curriculum and pedagogical decisions

- develop activities that promote observation and comparison skills that instill in students an appreciation for the value of life and the importance of caring for animals responsibly

- instruct students on safety precautions for handling live organisms and establish a plan for addressing issues such as allergies and fear of animals

- develop and implement a plan for future care or disposal of animals at the conclusion of the study as well as during school breaks and summer vacations

- espouse the importance of not conducting experimental procedures on animals if such procedures are likely to cause pain, induce nutritional deficiencies, or expose animals to parasites, hazardous/toxic chemicals, or radiation

- shelter animals when the classroom is being cleaned with chemical cleaners, sprayed with pesticides, and during other times when potentially harmful chemicals are being used

- refrain from releasing animals into a non-indigenous environment

A Word about Blood and Human Cells

Some teachers may recall a high school or middle school science class where they pricked their fingers to collect drops of blood for typing as part of a science investigation. Those days are long over. With blood-borne illnesses reaching near epidemic proportions, this act is too hazardous for students to continue. The risk of contamination is too great for even a marginal attempt. Instead, teachers can use artificial blood substitutes on the market for blood-typing activities.

Collecting human cheek cells for DNA extraction or for a slide stain is acceptable. However, with an increase in tuberculosis, teachers should not expose students to each other's saliva. A person with TB carries the bacterium in his or her saliva. If teaching in an area where TB is common, teachers are better off not allowing students to scrape their cheeks for cells for any activity. Anyone handling a toothpick or cotton swab that has been in someone's mouth could be exposed to TB or other respiratory tract disease-causing organisms. Instead, the teacher can conduct this lesson as a demonstration lesson, setting up microscopes with purchased slides or with his or her own cells between slides. This process becomes much easier if the class has access to a digital microscope that the teacher may plug into a computer to project onto a large screen.

Conclusion

This chapter focused on two big issues related to science instruction: acquiring equipment and supplies, and classroom safety. Although these topics may seem like separate ideas, they really are related. After all, how do teachers support students if they don't have the needed equipment or materials? Once they have the equipment and materials, they will definitely want to keep them, their students, and their schools safe when they use them. The overall message here is, science teachers can accomplish anything they want or need; they just need to dedicate time and attention into processes that come with protocols and follow through safely with their ideas.

Stop and Reflect

1. Brainstorm creative ways to obtain supplies for your science classroom.

2. What is the one piece of science equipment you wish you had? Find it online or in a catalog. Find a funding source. Write up a proposal or complete an application for this equipment. Submit your proposal or application.

3. What are some ways to convey to students the importance of safety in the science classroom?

4. What should you consider when using non-standard resources (e.g., chemicals, fire, electricity, animals, or blood or saliva)? When might you be inclined to use any one or more of these resources? Explain.

Chapter 4

What Is Inquiry?

"We have a hunger of the mind which asks for knowledge of all around us, and the more we gain, the more is our desire; the more we see, the more we are capable of seeing."

—Maria Mitchell

Most teachers would acknowledge the idea that students need to *do* science to *learn* science. However, not all science classrooms are filled with activity, research, and hands-on experiences. Teachers may avoid implementing inquiry-based instruction in their science classrooms for many reasons. It could be that it's too time consuming, too chaotic, classrooms are ill-equipped, or that teachers may not feel comfortable enough to let students explore ideas on their own. Preparing quality inquiry lessons is typically more time consuming for teachers than traditional textbook learning. Teachers need time to set up equipment and materials, organize their classrooms, and clean up. However, the results as evidenced by student learning are powerful. An inquiry-based classroom may be busier, nosier, and (sometimes) messier than traditional classrooms. Teachers should know that inquiry can be as structured or as open-ended as they allow, or that the topic allows. Scientific investigations can require the use of expensive materials and equipment, but they don't have to. Teachers should know that it is okay for students to explore and learn on their own. Teachers shouldn't feel like they have to be talking all the time for students to learn.

Defining Inquiry

The definition of *inquiry*, according to *Merriam-Webster Collegiate Dictionary* (2005) is: *1. examination into facts or principles; 2. a request for information; 3. A systematic investigation often a matter of public interest* (646). The *National Science Education Standards* define scientific inquiry as "the diverse ways in which scientists study the natural world and propose explanations based on the evidence derived from their work. Scientific inquiry also refers to the activities through which students develop knowledge and understanding of scientific ideas, as well as an understanding of how scientists study the natural world" (National Academy of Sciences 2013, 23). In essence, inquiry is a quest for knowledge. In science, this equates to an understanding of the natural world. Pretty much everything around us has to do with the natural world. Children are naturally curious, asking questions to seek understanding: *What makes a rainbow? Why do clouds float in the air? Why does my pillow spring back after I press on it, but a marshmallow does not? Why can't I see in the dark?*

Teachers have been conducting inquiry lessons in science for generations. Still today, if one were to ask any ten teachers what is meant by *scientific inquiry*, one could expect to hear ten very different ideas and interpretations on the topic. The Next Generation Science Standards, as mentioned in Chapter 1, include three dimensions. The first of these specifically addresses the idea of scientific practices. These practices clarify what students should know and be able to do as part of conducting a scientific inquiry.

> "… part of our intent in articulating the practices in Dimension 1 is to better specify what is meant by inquiry in science and the range of cognitive, social, and physical practices that it requires. As in all inquiry-based approaches to science teaching, our expectation is that students will themselves engage in the practices and not merely learn about them secondhand. Students cannot comprehend scientific practices, nor fully appreciate the nature of scientific knowledge itself, without directly experiencing those practices for themselves."
>
> —NGSS (2013f, 2)

If students are to "themselves engage in the practices," the role of the science teacher, then, is to provide opportunities for students to actively conduct investigations in which students actually "do" the science. This is best accomplished through inquiry. However, inquiry is more than just a set of practices. The *standards* inquiry is a step beyond the use of such skills as observation, inference, and experimentation. In addition, they require that students combine these practices with scientific knowledge (e.g., structures and properties of matter). This way, students use scientific reasoning and critical thinking to develop their understanding of the science content. The result is detailed in the Standards as performance expectations. These are the assessable statements detailing what students should know and be able to do after instruction. For example, a second-grade Next Generation Science Standard (2-PS1-1) reads "Plan and conduct an investigation to describe and classify different kinds of materials by their observable properties" (2013g, 13). In this instance, students must engage in the practice of "doing" science by planning and conducting an experiment. Then, they must perform an action: "describe and classify materials by their properties."

Science as Inquiry

Science as inquiry is basic to science education. It is a controlling principle in the teachers' planning of student activities. The Next Generation Science Standards use the term "practices" instead of "skills." Scientific practices allow students to develop an understanding about scientific inquiry as they participate in inquiry lessons. The standards are written as performance expectations. One part of these expectations is the practice of "doing" science. For example, students cannot demonstrate their proficiency in being able to analyze data without actually analyzing data. Students at all grade levels and in every domain of science should have the opportunity to use scientific inquiry and to develop the ability to think and act in ways associated with inquiry. This includes asking questions, planning and conducting investigations, using appropriate tools and techniques to gather data, thinking critically and logically about relationships between evidence and explanations, constructing and analyzing alternative explanations, and communicating scientific arguments. The science as inquiry standards are described in terms of activities resulting in student development of certain abilities and in terms of student understanding of inquiry. However, inquiry comes in many shapes and sizes, as explained in the following sections.

Confirmation Level Inquiry

Many "inquiry-based" activities that teachers may find online or in other resources are far from being true inquiries. Teacher demonstrations, although they do support inquiry-based instruction, are common activities. These types of inquiries generally confirm a scientific principle. Therefore, they are often referred to as *confirmation-level inquiries*. With a teacher demonstration, the teacher has one set of materials and equipment. He or she stands at the front of the class and walks students through the process of investigating. For example, to introduce the concept of sound energy, the teacher may bang on a pan close to a pile of rice placed on wax paper. The rice, responding to the sound energy created by the banging, jumps and moves. In this demonstration, the teacher is running the show entirely. The teacher sets up the situation, conducts the activity, and poses the questions. All the students do is watch and talk (and maybe write predictions or summarize what happened). This is a great activity to start students thinking about sound energy; however, it is not a true inquiry. Teacher demonstrations are quite time-friendly. They typically require few materials or equipment, but they engage students the least compared to other activities. Unfortunately, the cost-benefit to student learning is minimal. With a teacher demonstration, it is the teacher who does the bulk of the learning.

Structured Inquiry

Other common inquiry activities have students work in pairs or small groups to conduct an investigation. However, the outcome is predetermined, and the teacher usually sets up the investigation by providing the background information and the material resources for students to complete the predetermined steps. These are referred to as *structured inquiries*. For example, the teacher may pose the question, "Which loses its heat energy more quickly, a bathtub or a tea cup?" Each group of students may receive materials to conduct an experiment to answer this question, collect data, collaborate as a class to compile and compare data, and reach a conclusion. In this situation, the students are engaged in the work, but the teacher has guided them to complete a set of pre-determined steps. This activity started with a testable question. Students gathered data, analyzed it, and used it to formulate a conclusion. Many of the scientific practices were addressed with this investigation; however, the students could have just as easily rationalized the answer based on previous experiences. The experiment, such as it is, may really just confirm what students already know.

Guided Inquiry

One step beyond structured inquiry is *guided inquiry*. With this level of inquiry, the teacher typically poses a testable question to students. The next step, then, is for students to work to devise their own plan to reach a conclusion. Students take control of their own investigation, setting up the procedures, identifying needed materials, collecting and analyzing data, and ultimately reaching a conclusion. The teacher is there to support and guide students as they navigate their way through the inquiry process. But the inquiry is mostly student-generated. An example of guided inquiry would be to ask students, "How does the amount of sunlight affect plant growth?" or "How does the ground's surface affect bike speed?"

Open Inquiry

To truly be an *open inquiry*, students must start with a scientific question that they have generated, one that allows for research through an investigation, data analysis, and critical thinking. These are the kinds of investigations teachers hope to see during a science fair. With these investigations, students don't know the outcome. They make predictions based on research and prior knowledge. They design, conduct, and complete an investigation completely on their own. The teacher is there to guide their thinking and make sure they are on track to finish. All of the learning, from the initial inquiry to the final conclusion, is completely orchestrated and driven by the students. This is what true inquiry is all about. Possible questions may include, *Which home material makes the best sound insulator? What effects do power lines have on plant growth?* or *Do basketball shoes really help a person jump higher?* These are ideas students may be interested in, questions they may ask themselves, or an experiment they may design to find out something to which they don't know the answer.

Non-Inquiry Activities

Some "experiments" are anything but! Any activity that makes something or illustrates a concept is nothing more than a scientific model, or a *non-inquiry activity*. This includes having students create a model of a solar system, making a kaleidoscope, or cutting and gluing materials to a sheet of paper to show the parts of a flower. These are great activities. They are usually student-focused, and students enjoy doing them; however, they do not engage students in any of the eight scientific practices. There is little to no critical thinking involved. These types of activities do have a place in today's science classrooms, especially at the early primary grades but at a minimal level.

Planning for Inquiry

As illustrated, not all inquiry activities are the same. Some are more student-focused while others are more teacher-directed. In addition to the amount of student engagement versus teacher direction, inquiry activities may also vary in degrees of complexity. The more open-ended an investigation is, the more complex it becomes for students. However, open-ended investigations are the most rewarding for students, and they maximize the learning potential. Of course, teachers also must consider the time factor when designing and planning inquiry-based activities. Open-ended activities take the longest to complete, and teachers may feel as if they take up too much instructional time. Teachers need to consider all these factors, and have students participate in the most engaging activities that time and materials allow to maximize student learning.

Science by nature has a lot of inquiry. By recognizing the difference in the levels of inquiry, teachers may better plan activities to support student application of the eight mathematical practices from the Common Core Standards for Mathematics, and eventually lead them to be true inquirers. Figure 4.1 organizes the ideas described previously into the four levels of inquiry. This information is based on a summary by Heather Banchi and Randy Bell.

Figure 4.1 Levels of Scientific Inquiry

Level of Inquiry	Description and Example	Level of Teacher Direction	Level of Student Engagement
1 Confirmation	A scientific principle is confirmed. The results are known in advance. **Example:** Students model Earth's rotation and revolution.	High	Low
2 Structured Inquiry	The teacher presents a question which the students investigate following a set procedure. **Example:** Students drop various objects from the same height to see how gravity affects their fall.	Moderate	Moderate
3 Guided Inquiry	The teacher presents a question which the students investigate following a procedure they design and construct. **Example:** What factors (weight, length, or height of drop) affect the period of a pendulum?	Moderately Low	Moderately High
4 Open Inquiry	The students formulate a question and design procedures to collect data or evidence from which they may reach a reasonable conclusion. **Example:** Students design an experiment to test how different types of light (sun, incandescent, fluorescent) affect plant growth.	Low	High

(Adapted from Banchi and Bell 2008)

Science teachers in today's classrooms should quickly identify activities. These are more teacher-centered in nature. They typically don't require students to think a whole lot, or they are an end in themselves. True inquiries that ask testable questions, require data collection, analysis, and require creative thinking, lead to further investigations, are less teacher-centered, and insist that students put the eight science and engineering practices to good use. Students in today's science classrooms should be spending their time learning both science content and scientific processes.

Many activities available to science teachers may provide meaningful hands-on experiences for students, but they are not actually inquiry-based. Devising testable questions related to general science topics can be challenging. Figure 4.2 lists sample inquiry-based questions that students might investigate as part of the process of "doing" science. These can serve as models for teachers to follow when conducting inquiry activities in their classrooms.

Figure 4.2 Sample Testable Scientific Inquiries

Grade Range	Topic	Question
K–2	Earth, Sun and Moon	How does the temperature compare between day and night?
K–2	Plants	Which soil is best for growing sunflower seeds?
3–5	Rust	Which type of nails resist rust best?
3–5	Friction	Which surface creates the most friction with the bottom of my shoe?
6–8	Weather	How is humidity related to temperature?
6–8	Genetics	What is the relationship between the size of an animal and the size of its genome?
9–12	Ecology	How does the acidity of water affect the survival rate of brine shrimp?
9–12	Nutrition	How does the density of fruit affect its nutritional value?

The Scientific Method

The scientific method is a process by which students conduct a full experiment in their efforts to understand and explain something in the natural world. The scientific method is very closely related to inquiry. True inquiry and the scientific method both begin with a testable question. Additionally, they both require that students conduct a test, during which they collect data. They also both require an analysis of the data from which to form a conclusion. Teachers and students follow an organized set of procedures or steps to conduct an inquiry. We call these steps the scientific method. Publishers and reliable science resources vary in how they organize these steps. Sometimes they list six, seven, or eight steps. Although the procedures may vary slightly from author to author, the essence of the steps remains fairly consistent. Possible steps of the scientific method include:

1. Ask a testable question.

2. Conduct research about the topic.

3. Formulate a hypothesis or prediction about what will happen. Base this on prior experiences and the research.

4. Develop a procedure to follow and conduct the experiment. This includes a set of steps to follow, and materials and equipment needed.

5. Gather, organize, and analyze the data. Form a conclusion based on the evidence.

6. Communicate the results and identify other questions that arise from the test.

Example of a Structured Inquiry-Based Activity that Follows the Scientific Method

Question: What ratio of ingredients combine to make the best putty?

Research: Students conduct research on how putty is made and find various "recipes" to try.

Hypothesis: After researching this topic, students make an educated guess.

Materials:

- sealable plastic bag
- liquid white glue
- sodium borate
- water
- beakers
- stirs
- glass jars with lids
- protective gloves and eyewear
- permanent marker

Procedure:

1. Mix a 50/50 solution of glue and water. Use 200 mL of each. Place them in a glass jar. Close the lid and shake it until the water is completely gone. Use the marker. Label this jar as 50 percent solution.

2. Mix 10 mL sodium borate with 200 mL warm water in a second jar. Close the lid. Shake the jar until no particles of sodium borate remain. Use the marker. Label this jar 5 percent solution.

3. Mix these two solutions together in different ratios each in its own zipper bag to make putty.

4. Record the data.

5. Observe the properties of each putty. Decide on the one that has the best overall firmness, stickiness, and moisture.

Data Collection:

50% Glue Solution	5% Sodium Borate Solution	Observations	Physical Properties (color, texture, moisture content, stiffness)
15 mL	45 mL		
30 mL	30 mL		
45 mL	15 mL		
75 mL	15 mL		

Data Analysis: Students analyze their data, explaining what happened in each trial.

Conclusion: Students reach a conclusion based on the evidence they collected.

Further Questions: Students write additional testable questions that arose from conducting this experiment.

The "Nature of Science" Concepts

The skills, concepts, ideas, and practices related to following the scientific method are generally referred to as "nature of science" concepts. Appendix H of The Next Generation Science Standards (2013c) lists four nature of science understandings related to scientific practices.

- Scientific investigations use a variety of methods.
- Scientific knowledge is based on empirical evidence.
- Scientific knowledge is open to revision in light of new evidence.
- Scientific models, laws, mechanisms, and theories explain natural phenomena.

Teachers should consider the first bullet carefully. Although the scientific method is a common means of conducting an investigation, it is not the only method students might follow. Some questions may only be answered using research. For example, students could observe that they see a particular species of bird only during certain times of the year. Perhaps they see a particular species all year long, but only rarely. Students might ask, "What kind of bird is that, and why do I only see it in [spring, summer, fall, winter]?" Or, "Why do I see a bald eagle around my house? Is this bird within its range? If so, why aren't there more of them? If not, what brings it here?" Are these questions testable? Some might argue that yes, they are. Others might argue that no, they aren't. Are they inquiry-based? Students can use available resources to conduct research. They might also keep journals or logs of sightings: dates, times, locations. This data then becomes part of their evidence upon which they can draw a reasonable conclusion.

Remember that inquiry is a quest for knowledge about the natural world. Even though these questions do not naturally lead right into the scientific method, they do seek to learn something that presently is unknown. Students who ask these questions will not design any experiment, per se, but they will conduct an inquiry activity.

Experiments that follow the scientific method can take on many forms in the classroom. Teachers who teach young students may wish to conduct this type of investigation as a whole class. As the teacher guides students through the steps, individual students may participate in each step of the experiment. Or perhaps small groups or pairs of students may each have their own set of materials and supplies to "follow along" in a more hands-on fashion as the teacher leads the class. As students mature, teachers may allow students to conduct these types of inquiries in small groups, with less teacher guidance. Students might also conduct paired experiments. By the middle school level, and certainly by high school, students should have adequate independent learning skills to conduct experiments outside of class. This level of experimentation is reminiscent of a formal science fair.

Teachers in middle and high school should not assume that all students have the same level of independent work skills when it comes to creating and conducting a true inquiry following the scientific method. Some scaffolding may need to occur to help students be successful with this type of project.

This level of investigation might seem chaotic and disorganized to some teachers. Surely, if students are engaged in their own experiments at the same time in the classroom, they will be talking, conferring, moving around, and organizing their space and materials so as not to interfere with others. Groups may vary with regard to where they are in the process of completing their project at any point in time. While some students are researching, others may be following their procedures, while others still may be wrapping up their project and finalizing their report. Teachers can help maintain order in a busy science classroom by following some of the suggestions listed later in this chapter.

Science Technology Engineering and Mathematics (STEM) Activities

Recent studies indicate that students in the United States lag behind their international peers in STEM-related interests. A report to the President by the President's Council of Advisors on Science and Technology (2010) summarizes the following information:

- Less than one-third of eighth graders are proficient in math and science.

- About 33 percent of bachelor's degrees earned in the United States are in a STEM-related field. Compare this to the average percentage in China at over 50 percent, and in Japan at over 60 percent.

- Over half of the science and engineering graduate students attending universities in the United States are from other countries.

These statistics are concerning. According to an Executive Summary published by the United States Department of Commerce (2011), jobs in STEM fields have grown three times as fast as jobs in non-STEM fields. This same report has predicted STEM field careers to grow by 17 percent between 2008 and 2018, compared with just a 9.8 percent growth for non-STEM jobs.

The federal government has supported STEM programs financially. Agencies, industries, and organizations are also interested in improving students' interest in STEM careers. These include the STEM Education Coalition and STEM schools, such as Edison Collegiate High School in Ft. Myers, Florida and Thomas Jefferson High School for Science and Technology in Alexandria, Viginia.

STEM is an initiative that promotes both mathematics and science through creative problem solving. When students participate in a STEM activity, they are presented with a real-world problem or situation. They use science knowledge and engineering practices to devise a solution to the problem by implementing the following engineering design process:

1. Identify the problem or need.

2. Conduct background research.

3. Plan one or more solutions to the problem or need.

4. Decide on one solution to investigate.

5. Create a model or prototype.

6. Test the model or prototype. Analyze what worked, what didn't, and why. Determine how the model or prototype might be improved.

7. Redesign the model or prototype. Repeat step 6 until it reaches the desired results.

The following is an example of a STEM activity:

Situation: A local Parks and Recreation project manager has requested permission to build a new park in a centrally located area of town. Citizens in this area feel the center of town to be an undesirable location. They are concerned about costs related to continued park maintenance, lost tax revenue (county land is tax exempt in this area), and the potential loss of employment (once the park is completed, no business may move into this area).

Task: Design a plan for a park in a fictional town. Include in the plan an area for conservation of natural resources and wildlife, a revenue-generating system for the immediate community, and a means for local citizens to come together as a community for a celebratory event. The park must contain elements to address each of these areas.

Conduct Research: Find out how parks benefit local communities with regard to ecology, economics, and citizenship.

Develop a Plan: Create a blueprint of the park. Identify key areas using a coding system.

Communicate Results: Develop a complete plan, in writing, to explain in detail how this park addresses all of the citizens' concerns.

STEM activities fall at the highest level of inquiry. Students complete most or all of the work, with guidance and support from their teachers as they work their way through the entire process. These are very closely related to Level 4 science inquiries. In Level 4 inquiries, students conduct an investigation for a testable question. In STEM activities, students apply this learning for a real-life purpose, suited to address a real-life problem. They actually design and construct (or plan) a model to address the problem, and test its effectiveness. STEM activities generate learning environments very similar to those where the scientific method is in full swing. Depending on the activity, students may be designing prototypes using various materials and equipment. They likely are collaborating to complete their project, which requires talking and movement.

Inquiry and Technology

Many students are highly interested in using computers to investigate science phenomena. Today's science classrooms likely have technology resources that can aide teachers in the higher levels of scientific inquiry. Online simulations, games, equipment, and apps extend the four walls of the science classroom and allow students to participate in activities that they otherwise would not be able to do. For example, students can compare onion skin or cheek cells using online microscopes. They can participate in a virtual frog, pig, or worm dissection. They can even observe a virtual DNA extraction. Younger students can take a virtual field trip to the zoo, or grow virtual plants using online resources. Students of all ages can participate in online games and simulations that challenge their knowledge and require creative thinking as they create environments based on the weather, decide whether objects will sink or float, or design and test a rollercoaster.

Many appropriate simulated labs are widely available on the Internet. When teachers cannot adequately provide the proper material for a hands-on investigation, a virtual reality investigation can act as a sound substitute. Teachers and students can go online to collaborate with another class, school, or group on a long-range project. Learning apps for mobile devices allow today's science teachers to really explore the vast world of science. These apps allow users to access information about the periodic table of elements, make scientific calculations, or learn about human anatomy. Online magazines such as *Focus Magazine* and *Scientific American* list today's top science apps for students and teachers. Much of this application software supports the inquiries that students may be conducting. Imagine the level of student engagement in a science class where students are actually encouraged to pull out their smartphones and use appropriate application software for educational purposes.

Online resources aside, teachers might also be able to use digital equipment with students to conduct inquiry activities. In addition to digital science equipment, teachers might also have access to interactive whiteboards—the perfect medium through which to access and use online simulations and games with students.

The integration of technology is only limited to how today's science teachers can dream to use it. Video chatting and texting, social networking, and blogging might be students' favorite means of staying connected to their friends and family. Science teachers, too, can set up their own accounts and have students contribute to them as their personal (or school-based) technology will allow. As far as using technology for introducing, participating in, and exploring inquiry-based activities, teachers should not overlook these varied, creative, and engaging resources.

Classroom Tips for Using Inquiry-Based Instruction

Classrooms in school buildings may lack the square footage necessary or desired for science teachers to organize themselves and their students as well as their equipment, materials, and resources safely. Indeed, inquiry-based classrooms require that students work collaboratively at work stations and have access to the equipment and materials they need. Depending on the learning environment, some teachers may need to get creative with the space and resources they have in order to maximize students' involvement in inquiry activities. The following are a few tips to help teachers get started.

Organizing Space

For inquiry-based activities, students should work in groups of three or four. If there are any more than that, then someone tends to get left out of the hands-on experience. Students should be grouped together. If desks are situated in traditional rows, students can pull them together to form mini-circles or mini-stations. If students are seated at two-person stations, one pair may partner with the pair behind them. In this case, all that is required as far as furniture shifting is the moving of chairs, not desks or tables.

If inquiry activities require movement or exploration outside the classroom, students can still work in small groups. The teacher should provide directions and distribute materials inside the classroom so that once students exit the room, they may begin working right away. Some activities may require large amounts of space. In this instance, teachers may consider using the gym or cafeteria, depending on availability. The school may have an open area such as a quad that can provide adequate work space for students or a hallway or covered walkway may suffice.

Organizing People

Teachers need to plan how they wish to organize students prior to beginning inquiry activities. Some activities may be best completed independently, some, in pairs, while others may require lab groups. Deciding who works with whom is an essential preparation for teachers of all grade levels. Pre-determined groups ensure everyone is accounted for, and no one gets left out. Groups should be balanced so that each member has the opportunity to share in the responsibility of the activity.

Teachers do not want one student in a group doing all the work while the others sit back and watch. According to Stanford University, "Individual accountability is essential to group success, since the natural tendencies of some students to dominate and some to withdraw will gradually come into play unless some mechanism is in place requiring everyone to participate" (1999, 2). To alleviate these situations, teachers can assign permanent or semi-permanent lab groups, and have each student in the group take on a collaborative role. These roles hold all students accountable for the group's work, and each student is active in all the learning that takes place. Some popular group roles include: reader, recorder, clarifier, summarizer, taskmaster, materials engineer, and encourager. Teachers may assign these or similar roles based on students' strengths and give them the opportunity to rotate these roles as the year goes on. With enough inquiry-based learning occurring, students would likely serve several roles over the course of a year, thus promoting a true collaborative effort.

If group work is new or challenging for some students or some classes, teachers may need to take time to "teach" students how to work collaboratively. The teacher and the students can set the ground rules or expectations for group activities. These should be clearly written, posted, and reviewed before any work begins. Students who do not adhere to the expectations should not expect to continue working. Investigations are usually engaging and interesting and students don't want to miss them. By setting clear expectations and following through with reasonable consequences when not followed, teachers and students can enjoy productive activities all year long.

Suggested Expectations for Group Work

- Speak cordially with each other
- Share materials
- Give everyone a turn
- Listen respectfully to everyone's ideas
- Treat the equipment with respect
- Have everyone pitch in to clean up

Organizing Materials and Equipment

Another consideration with regard to the overall classroom construct is the organization of equipment, supplies, and lab materials. "The classroom environment is influenced by the guidelines established for its operation, its users, and its physical elements" (Stronge, Tucker, and Hindman 2004, 2). Some classrooms may have adequate space, and the activity may allow for students to acquire the equipment and supplies from one central area to use at their seats. In this instance, teachers may consider having trays organized with the necessary materials. Then, one student from each group can retrieve the necessary components in one trip. This way, students will not lose class time waiting for the teacher to count out craft sticks, measure soil, or rip off plastic wrap. If using equipment that requires electricity, such as microscopes or hot plates, keeping these items stationary and rotating student groups to them may be a better organizational strategy. While one or two groups are at the microscopes along the counter or table, the other groups may be discussing the experiment, reading, or researching information about the topic. Then, when there is room at the counters, the groups trade places. Those who were reading now use the equipment, and vice versa.

Some schools may have the benefit of a portable lab. These labs may have computers, hand-held devices, as well as other instructional technology such as scientific graphing calculators, CO_2 sensors, or high-powered microscopes. In addition, labs may include special materials such as litmus paper, soil samples, petri dishes, or cultures. They may also house ready-made labs, complete with instructions and materials for several student groups, or specialty items such as an inflatable planetarium or a working water cycle model. In tight quarters, teachers may need to move desks aside, have students sit in chairs in a semicircle, and use clipboards upon which to write (if students do not have hand-held devices at their disposal).

Time Management

Before it all begins, teachers need to spend time preparing the investigation for their students. This requires preparation time outside of student contact time, either before or after school, the day before the activity. Teachers should have the necessary equipment, materials, and supplies ready and waiting for students when they arrive to class. Teachers should budget adequate cleanup time after the investigation is complete or before class ends each day. To help with this, teachers can set a timer to sound when five or ten minutes are left of class. Once sounded, all group work should cease immediately, and cleanup should begin. Teachers might also want to warn students at the halfway point of an activity, and encourage groups to move along if they lag behind others. Time management really requires that teachers be prepared and have a vision as to how they want activities to run before students start working.

Inquiry-based activities can be quite time-consuming. Some can be completed in one class period; others may need to be repeated or completed over several class periods. Teachers should know or have an idea as to how long each activity should take. Even the most prepared teachers may discover that a particular activity requires more or less time than anticipated. Teachers should understand that they may need to alter their daily or weekly plans, depending on how the lesson goes. McLeod, Fisher, and Hoover state "Teachers who effectively manage time give their students the best opportunity to learn and to develop personal habits that lead to wise use of time" (2003, 9). Using time management in the classroom effectively will lead to student success!

Conclusion

All science roads lead to inquiry. However, teachers should understand what are true inquiry-based activities. Level 4 activities tend to be the most time-consuming, but they are the most engaging to students, leading to deeper learning outcomes. Most true inquiries follow the scientific method. However, this is just one means of addressing an inquiry-based question. Students may simply engage in a research project to answer a question about the natural world, or they may participate in a challenging, meaningful STEM activity in an attempt to solve a real-world problem or situation.

Inquiry comes to science in all shapes and sizes. Teachers can put technology to good use investigating ideas that might otherwise go unexplained. Additionally, inquiry requires careful and thoughtful planning and preparation. When teachers give careful consideration to the organization of people, materials, resources, and time, inquiry-based activities run more smoothly, and students can focus on gaining greater insights to the world around them.

Stop and Reflect

1. How do the Next Generation Science Standards support inquiry-based instruction?

2. Describe one confirmation or structured inquiry lesson you recently have completed with students. Now, consider how you might revise it to be more guided or open-ended.

3. What are the greatest roadblocks to implementing more guided and open-ended inquiry activities with your students? How might these roadblocks be overcome?

4. Find an online simulation or other student-centered technology to use during an upcoming unit. Record the name of the activity and where you found it. Note how and during which stage of the learning process (beginning, middle, or end) you plan to use it.

Chapter 5

Effective Instructional Design

"How did we get to the point where teachers hope for good results rather than plan for them?"

—Jane E. Pollock

There is no mystery to developing an effective science lesson. The lesson plan should follow the general outline of any other lesson plan in any subject area. Of course, teacher strengths, classroom climate, and student interests play key roles in whether students meet the outcomes set forth in the lesson. However, with careful and deliberate planning, every science lesson can reach for, meet, and even exceed the teacher's expectations for their students.

Begin with the Standards

Whether teachers follow the national, state, or district science standards, all standards-based instruction begins here. Teachers may rely on their science series to lay out a year-long scope and sequence of science learning for students. This resource is useful and viable, especially for beginning teachers who may need a more structured outline to follow.

Plan with Resources

Science teachers should take heed before entirely depending on a published program to provide appropriate, relevant, and necessary learning opportunities for all students. Project 2061 was founded in 1985 by the American Association for the Advancement of Science (AAAS). Named for the year in which Halley's Comet will return, it is a long-term initiative to help all Americans become literate in science, mathematics, and technology. One aspect of Project 2061 is textbook evaluations. Middle and high school mathematics and science textbooks have been analyzed and evaluated to determine how well they meet specific criteria important to advancing science and math literacy among United States students. Beyond content, one of the most important criteria was the level of instructional support provided to the teacher. Of the ten middle school science textbooks evaluated for quality of instructional support in the United States, not one was rated even "satisfactory," according to AAAS criteria (Koppal 2002). These textbooks included multiple classroom activities that were either irrelevant to learning key science ideas or didn't help students relate what they were doing to the topics they were studying. High school biology textbooks fared slightly better.

In addition to lack of instructional support, many textbooks also have some factual errors, particularly in drawings and other diagrams. They are usually representations that are likely either to give rise to or reinforce misconceptions commonly held by students (Koppal 2002). In one of these texts, diagrams that show the photosynthesis and respiration relationship between animals and plants can be confusing. These diagrams fail to clearly show that plants respire, just like animals. They also tend to give false information. For example, one textbook noted that photosynthesis and respiration balance one another, instead of the fact that the rate of photosynthesis is beyond that of respiration. When this is stated correctly, students can understand that plants produce enough food during photosynthesis to meet both their own needs and the needs of other organisms (Koppal 2002). Another example is a common claim in elementary textbooks that clouds remain aloft because the water droplets are so small and widely spaced that gravity has less effect on them, but the total mass of the water is the same whether it is all together in a pool or spaced out in a cloud. Gravity should pull it down. Why does it stay aloft? The answer is that the mass of the cloud's droplets is buoyed by the upward movement of the heated air between the droplets. As water vapor (gas) condenses into water droplets (liquid), it gives off heat (a phase change). This heated air expands and becomes less dense. As it becomes less dense, it moves upward in the atmosphere. Clouds stay up in the sky because, on average, they are much less dense than the surrounding air. Once the water droplets reach a critical mass and fall out of the clouds as rain, the remaining hot air is no longer weighed down by tons and tons of water, and it races upward. This rising hot air creates the violent updrafts in thunderstorms and hurricanes. There are many such examples of errors that give rise to misconceptions in elementary and middle school science textbooks. Therefore, a science textbook should be neither the sole nor the main source of information or instructional resources in the classroom. Teachers need to first understand the science concepts they are required to teach. Then, they should become aware of common student misconceptions regarding these concepts. Teachers should also be on the lookout for common errors in textbooks. Once these formative processes have taken place, teachers may work to build instructional plans to meet the content objectives, clarify students' misunderstandings or misconceptions of the content, and provide accurate, student-friendly informational text upon which to build students' content knowledge.

Consider Lesson Objectives

"What should the student know and/or be able to do by the end of this lesson/unit?" This question leads teachers to identify and write specific learning objectives based on the science standards they are using to guide instruction. These are different from learning goals. Goals are generally broad and immeasurable. Objectives, in contrast, are measurable. They answer the question posed at the beginning of this paragraph. Examples of comparing goals from the Next Generation Science Standards (2013g) and objectives can be found in Figure 5.1.

Figure 5.1 Examples of Goals vs. Learning Objectives

Grade-Level Range	Goal	Objective
1–2	Students will recognize that offspring resemble their parents.	Students will correctly match five out of five pictures of offspring with pictures of their parents.
5–6	Students will understand how plants and animals adapt to live in their environment.	Students will identify three desert plants or animals and give two or more details of each to explain how it adapts to survive in the desert.
8–9	Students will understand that the diversity of life on Earth is always changing.	Students will give at least two reasons to explain, in detail, why environments with high biodiversity are more stable than environments with low biodiversity.

Learning objectives are important because they establish the final destination students should reach on their learning journey, or instructional plan. Teachers may easily access information, activities, ideas, and learning strategies in this digital world. The teacher must set a clear objective or set of objectives for his or her students. Look at the fifth/sixth grade example in Figure 5.1. Where does one start? There are so many environments and so many animals with adaptations that allow them to live there. One can see how easily a teacher and his or her students might go down too many paths and never reach the goal. For example, a teacher might have students tape their thumbs to their hands and attempt to complete everyday activities (e.g., tie their shoes); analyze a graph that compares various animal births to survival rates; and relate the terms population, community, competition, adaptation, and symbiosis in any three or more environments. These are all valuable activities, and students should learn from them; however, they do not necessarily provide the means for students to demonstrate their understanding of plant and animal adaptations as stated in the learning objective. Instead, more appropriate activities might be to compare the evaporation rates of a wet sponge covered in petroleum jelly and one without, analyze the physical characteristics of desert animals (by watching short films or reading informational text about them), and sort plants and animals by the characteristics that enable them to survive in a desert climate. These activities help the teacher and student stay focused and allow students to build on their existing knowledge about plants and animals (and where they live), conceptualize the idea of desert dwellers' adaptations, and demonstrate their understanding of desert plant and animal adaptations. The students engage in an activity directly related to the objective by learning about specific adaptations, and confirming their understanding of these adaptations. Then, once students have a solid foundation of how desert adaptations help plants and animals survive, they may move onto more complex ecosystems, and begin to compare and contrast the adaptations of plants and animals from other environments to those of the desert. This learning plan is much more sequential, organized, and meaningful. It allows students to focus on one topic, then apply their knowledge to other closely related topics.

Move Toward Higher-Level Outcomes

As teachers become accustomed to writing effective learning objectives, they can also begin moving students toward higher-level outcomes. For example, if students can successfully explain how animals have adapted to live in the desert, they should have a general understanding of the idea of animal adaptations. However, are they able to apply this knowledge in a variety of contexts? Another outcome may be that students design a new species that has uniquely adapted to the desert environment, and write a placard for the zoo where this animal is kept to explain where it lives and its adaptations for living there. In order for students to achieve this particular outcome, they must not only know and understand plant and animal adaptations, but they must also synthesize this information in a unique and novel manner. This level of learning requires students to think analytically, critically, and conceptually, which are all higher levels of Bloom's Taxonomy. These are all important skills if science teachers are concerned with developing students who not only know facts about science but can also apply them to real-world situations. This idea falls into the science and engineering practices of the Next Generation Science Standards (2013g). This type of activity requires students to *apply* content knowledge, not just *have* content knowledge.

Consider Student Outcomes

Closely related to the idea of writing clear, concise, measurable learning objectives is how students will go about demonstrating their understanding. The objectives answer the question "What will students know and/or be able to do by the end of this lesson/unit?" Student outcomes determine how students will demonstrate their knowledge and/or skill related to the objective. In other words, what will be the students' evidence of learning? Teachers may have students answer a multiple-choice test about a topic, or correctly use related vocabulary terms in a cloze-like activity. These are just a few methods teachers may use to assess whether students have mastered the learning objective(s). Figure 5.2 includes several other suggestions beyond the paper-pencil (or digital) test. To evaluate students' mastery of the objectives using these types of tasks, teachers may use evaluation checklists or rubrics.

> A cloze activity requires students to fill in sentences with appropriate vocabulary terms. Teachers use this method of vocabulary assessment to determine whether students can correctly use terms in context.

Figure 5.2 Examples of Student Outcomes (Evidence of Learning)

- cartoons
- concept maps
- essays
- fictional interviews (between scientists or organisms)
- game show questions and answers
- multimedia presentations
- one-sentence summaries (or other summaries)
- science journals
- scrapbooks
- songs

In addition to the *how* of demonstrating understanding, teachers should also consider the manner in which this assessment will take place. Will students work independently, in pairs, or in a small group? Depending on the objective(s), a teacher may want students to work collaboratively. The teacher may make note of student mastery/non-mastery using an observation checklist. For example, if the learning objective is that students correctly use a triple beam balance scale, the teacher, by monitoring small group work, can easily see who can and who can't. Then, during future learning activities that require the use of this equipment, the teacher can demonstrate for the student who is having difficulty with this skill and monitor his or her attempts more closely until mastery is determined.

Lesson Formats

The Next Generation Science Standards clearly state no strategy, approach, or method of teaching science should be dictated when designing lessons (NGSS 2013a). They acknowledge that students learn through various modalities, respond to varied techniques, and bring their own set of conceptions to each science class. Therefore, no one manner of teaching science is appropriate in all instructional settings. This provides teachers the opportunity to develop and create their own lesson plans and to modify plans to better meet students' learning styles.

One popular method for creating lesson plans is called the 5E instructional model. This model has a long history in education. It is related to constructivism, reflective thinking, and character development (Chitman-Booker and Kopp 2013). The 5E was formally conceptualized by the Biological Studies Curriculum Study (BSCS) in the mid-1980s. This instructional model provides a structured, balanced approach to teaching science that all teachers can use and modify depending on the content and the students. A summary of the steps to the 5E instructional model can be found in Figure 5.3.

Figure 5.3 Steps to the 5E Instructional Model

5E Step	Description	Example
Engage	The "hook"; a short introduction to the topic	Ask, "Why don't monkeys run around your house instead of squirrels?"
Explore	Hands-on experiences for students	Students match labeled pictures of animals to the environment in which they think they live (e.g., desert, arctic, rain forest, ocean).
Explain	Content defined and explained	Students watch videos and take notes of animals in their environment, and they read informational text about animals in their environment. Students record this information in a matrix for each environment.
Elaborate or Extend	An extension of what students have learned	With a partner, students research various animals that live in different parts of the world. They identify the characteristics that allow the animal to survive there, including its covering and defense against predators.
Evaluate	Assessment of student learning	Students design and illustrate an original animal suited to live in a particular environment. They write a summary explaining how it is adapted to live there. These summaries will be compiled into one class book of "Unique Creatures of the Wild."

Many teachers see the value in following a 5E science lesson plan format. It includes elements from various teaching methods, particularly inquiry-based instruction, traditional textbook or lecture-style teaching, and collaborative learning. Depending on the nature of the science concept, the class time allowed, and the background knowledge students bring to the science classroom, teachers may develop lesson plans that follow one method for a period of time, then move onto a different method. Furthermore, teachers will need to decide how to organize the students. Will they work independently or in groups? Will the teacher engage them in a class discussion, or will he or she ask pairs or small groups of students to discuss a topic? Each classroom teaching and learning style is unique, therefore, the lesson plan format must be flexible enough to maximize learning in the setting, time, and space allowed. Teachers have many options when it comes to designing effective lesson plans. They may follow any one of several lesson plan formats. Figure 5.4 lists several options for teachers. For teachers who may tend to lean toward lecture, they may see an option that might re-energize their classrooms. For teachers who tend to lean toward inquiry, they may see an option that might strengthen reading-to-learn opportunities for their students.

Figure 5.4 Suggested Lesson Formats

Lecture Models	Inquiry-Based Models	Collaborative Models
• Slideshow presentations • Teacher-led class discussions • Questions and answers • Teacher demonstrations • Guest speakers • Textbook/reading assignments • Interviews with notable or community scientists	• Labs • Investigations • Hands-on activities • Student-led demonstrations • Field work • Simulations • Field trips	• Discussion groups • Student-led presentations • Research reports • Debates • Socratic method • Games • Online learning • Drama/role playing • Problem solving/case studies • Peer teaching • Reciprocal teaching

Differentiating Instruction

Academic background knowledge is one leading indicator of students' academic achievement (Marzano 2004). Prior knowledge and background experiences of students are elements that are out of teachers' hands. For example, for students to understand the concepts of evaporation and condensation, they need to have a solid understanding of the states of matter. Also, students with learning disabilities and English language learners may struggle to understand some concepts. The information may not make sense to them. They may need repeated instruction to demonstrate even a modicum of understanding. Lack of adequate vocabulary skills, particularly with regard to content-specific vocabulary, may compound students' difficulties with science topics. Conversely, science teachers may discover that some students in their class already have a solid understanding of the topic or concept. Therefore, science teachers need to differentiate instruction for these students.

One way teachers may learn what knowledge and experiences students bring to a particular topic is to administer a pre-test or other initial "screening" assessment. Once reviewed, teachers can quickly identify information students are missing that is essential to their understanding of the upcoming topic. Teachers could develop a short questionnaire for students regarding information about an upcoming topic. This will help students to begin thinking about the topic and assess their prior knowledge before beginning instruction. Teachers might also provide students with a scenario related to a concept and have students respond to a series of questions. For example, a middle school teacher might ask his or her students what they think might happen if an astronaut's tether were to be severed. After the students' responses are shared, the teacher can record notes about students' misconceptions about force and motion. Then, he or she may include additional learning opportunities to correct these misconceptions. A teacher may also use a KWL chart as a precursor activity (what I *know*, what I *want* to know, and what I *learned*). This activity provides teachers with information about both students' background experiences and misconceptions. Students complete the *K* and *W* columns before learning. Then, they complete the *L* column after learning, clarifying and correcting original ideas as they go. Figure 5.5 provides an example of a KWL chart on the topic of how plants and animals compare.

Figure 5.5 Sample KWL Chart

What I KNOW	What I WANT to Know	What I LEARNED
• Some animals have fur. • Plants don't have teeth. • Animals have four legs. • Plants have roots. • You can eat most plants and animals. • Plants have seeds. • Animals have babies. • Many plants are green.	• Why are some plants poisonous? • Why do I have to eat vegetables? • Why do animals have claws? • What's the biggest animal? • What's the smallest animal? • Can plants grow in snow? • Are there plants in outer space?	• Animals have fur, claws, and teeth to protect themselves. • Some plants protect themselves from being eaten by making poisons in their leaves. • Plants need air, water, and sunlight to grow. • Vegetables are important for good health. • The biggest land animal is the African elephant. • The biggest animal of all is the blue whale.

Once a teacher has identified students' misconceptions or lack of essential background knowledge, he or she may begin infusing differentiation strategies into the daily and/or weekly lesson plans. Differentiation models and strategies are greatly varied and quite plentiful in today's educational climate. Some suggestions for differentiating instruction are listed in the following box. They range from simple modifications to the lesson model to more time-consuming preparations of instructional materials. However, once teachers have completed any material modifications, they may be saved and used during subsequent years. If teachers can collaborate with others to create these materials, they share the workload and make their time investment go farther.

Differentiation Suggestions

- Use collaborative learning strategies.

- Use hands-on experiences as often as possible.

- Have students create concept maps of topics using essential vocabulary terms.

- Provide diagrams and illustrations of processes that students can label.

- Provide students with an outline of the text.

- Provide students with leveled text better matched to their reading abilities.

- Provide students with graphic organizers matched to the text structures: cause-effect, main idea and details, sequencing, compare and contrast, etc.

- Provide students with pre-typed notes on the subject matter. Students may jot reflections or connections to the ideas in the margins.

Integrating Technology into Instruction

Today's science classrooms have the benefit of using technology during instruction. The Internet, hand-held devices, classroom computers, video and digital photo devices, recording equipment, interactive whiteboards, and advanced science equipment can be welcomed contributions to a teacher's lesson plans. They can enhance the learning experience by engaging students in what they know and love—the digital world. Teachers can use available technology for any number of lessons, including, but not limited to online simulations and demonstrations, instructional videos, online activities and games, social networking, and publication and presentation software.

This use of technology opens a whole new world of lesson plan development for science teachers, as it provides each student with his or her own hand-held device or computer. In this classroom setting, students use the computer almost exclusively for learning purposes. Many schools and now many school districts are headed toward implementing one-to-one programs. One example is Sunnyside Unified School District (a tech savvy school district) in Tucson, AZ. They state, "At Sunnyside, the one-to-one program design is a learning environment in which every student, grades 4–12, is provided a laptop on a direct and continuous basis throughout the school day and beyond" (Sunnyside Unified School District 2014, under "District").

Some teachers may embrace this paradigm shift; others may be fearful of it. If a school or district chooses to engage in one-to-one, the science team can collaborate to discuss the best use of this instructional model and the technology afforded to them. One idea is called a "flipped" classroom. In this model, the teacher sets up a learning activity such as an instructional video or reading assignment that students watch or complete at home. Then, in class, the teacher, having already "taught" the lesson digitally, can provide students with more hands-on and simulated learning activities. Bryan Goodwin and Kirsten Miller explain the model like this: "Some teachers are now creating flipped or inverted classrooms in which they record lectures and post them online. Students watch the lectures at home, where they can speed through content they already understand or stop and review content they missed the first time the teacher discussed it (and might have been too embarrassed to ask their teacher to repeat in class). Online lectures can also easily incorporate visual representations, such as interactive graphs, videos, or photos of important historical events" (Goodwin and Miller 2013, under "Home"). Since the lesson was really homework, this frees up class time to engage students more fully in the science activity. The one-to-one initiative is very exciting for today's science teachers. They will be able to accomplish much more in the time they have with students during the school day.

Conclusion

Good instructional design begins with the national, state, or district science standards. These provide the destinations for student learning. To develop an effective lesson plan, teachers should first carefully consider how these standards translate into learning objectives. Then, teachers will need to decide how they want students to demonstrate their understanding. Finally, teachers may plan and prepare a focused series of instructional learning opportunities (hands-on, text-based, or simulated) to help students reach the goal.

Stop and Reflect

1. On a scale of 1 to 5 (1 = not at all; 3 = somewhat; 5 = all in), how would you rate your lesson planning style as standards-based? Justify your evaluation.

2. How do you see exploration before content helpful as students begin to learn new science information?

3. Write one to three lesson or unit objectives for an upcoming science unit. Then, design assessment tools to help students demonstrate mastery of this/ these objective(s).

4. Think of one student who seems to struggle with science content. What differentiation strategies can you use to support this student with a current topic of study?

Chapter 6

Literacy in Science

 "All that mankind has done, thought, gained, or been; it is lying as in magic preservation in the pages of books."

—Thomas Carlyle

Middle and high school science teachers typically identify themselves as just that: science teachers. However, in the elementary setting, there may not be time set aside specifically for science instruction. In today's classrooms, one observer may see a blended curriculum, with science teachers integrating reading and writing, and literacy teachers using science-related informational text to teach literacy skills. Since science textbooks are heavily loaded with difficult vocabulary words and complex concepts, it makes sense to encourage students to write about what they read, as it can help solidify their learning. Writing provides a bridge between the content knowledge and understanding (Michigan Science Teachers Association 2014). It makes sense for teachers to collaborate to help students become not just literate but scientifically literate as well. The integration of science in the literacy block, and/or literacy in the science block, may help teachers accomplish both of these objectives.

The relationship between reading and writing cannot be ignored. Simply put, Joelle Brummit-Yale states, "reading effects writing and writing effects reading" (K–12 Reader 2014, under "The Relationship Between Reading and Writing"). For example, reading a plethora of literary genres helps students learn text structure and phrasing that can be applied to their writing. Conversely, students can utilize the phonemic awareness skills they have developed while reading and apply those skills to their writing. Lastly, skills that students use in the writing process such as peer editing

Appendix M of the Next Generation Science Standards is titled "Connections to the Common Core State Standards for Literacy in Science and Technical Subjects." It provides an alignment tool between the science and engineering practices and the corresponding literacy standards.

and author's chair help them learn to read. These questions can include *What did they mean when they said _____?* or *How could you say that in a different way?* In short, they are constantly thinking about what the story is about and ways to improve their writing. These are the types of questions students should be thinking about as they critically read a text.

Reading Standards in Science

With the onset of the Common Core State Standards in English Language Arts Standards came the requirement for teachers to integrate nonfiction text into their curriculum. This is good news for teachers who only teach science. This means that students should be coming to class better prepared to read what can at times be very difficult nonfiction text. But the road to literate students cannot only befall the shoulders of the English language arts teacher. Science teachers can help students learn and apply the literacy skills they learn in science class. Specifically, students need to learn to read informational text. The Common Core English Language Arts Standards refer to these as Reading for Information (RI) Standards. They are listed in Figure 6.1.

Figure 6.1 CCSS ELA Reading for Information Anchor Standards

Common Core Reading for Information Standards	
CCRA.R.1	Read closely to determine what the text says explicitly and to make logical inferences from it; cite specific textual evidence when writing or speaking to support conclusions drawn from the text.
CCRA.R.2	Determine central ideas or themes of a text and analyze their development; summarize the key supporting details and ideas.
CCRA.R.3	Analyze how and why individuals, events, and ideas develop and interact over the course of a text.
CCRA.R.4	Interpret words and phrases as they are used in a text, including determining technical, connotative, and figurative meanings, and analyze how specific word choices shape meaning or tone.
CCRA.R.5	Analyze the structure of texts, including how specific sentences, paragraphs, and larger portions of the text (e.g., a section, chapter, scene, or stanza) relate to each other and the whole.
CCRA.R.6	Assess how point of view or purpose shapes the content and style of text.
CCRA.R.7	Integrate and evaluate content presented in diverse formats and media, including visually and quantitatively, as well as in words.
CCRA.R.8	Delineate and evaluate the argument and specific claims in a text, including the validity of the reasoning as well as the relevance and sufficiency of the evidence.
CCRA.R.9	Analyze how two or more texts address similar themes or topics in order to build knowledge or to compare the approaches the authors take.
CCRA.R.10	Read and comprehend complex literary and informational text independently and proficiently.

—CCSSO (2010, 10)

As a science teacher, consider how these literacy standards can support students as they learn the information they need in order to proficiently learn the Disciplinary Core Ideas, one of the three dimensions of the Next Generation Science Standards (2013f) as discussed in Chapter 1. For example, English Language Arts Standard 6 (2010) requires students to consider information presented from two points of view. How much learning would take place if students were to read a first-hand account of an astronaut's experiences in space? Then compare these ideas to those presented in nonfiction text about space travel, but written from a third-person perspective (e.g., informational article or trade book)? Look closely at Standard 7. This particular standard can easily be written for the science classroom. Students can compare information presented only visually and in words, such as in a diagram of the nitrogen cycle, and compare it to a text-only summary of the nitrogen cycle. What might students learn from looking at a chart summarizing Mohs scale of mineral hardness, then comparing this information to a text-only version of this topic?

Many types of nonfiction text support science teachers with the integration of these Reading for Information Standards. These include, but are not limited to, nonfiction trade books, textbooks, reference books, periodicals, journals, newspapers, fliers, brochures, and reputable informational websites. Science teachers may easily liven up the text students read by obtaining information from a local county or state park, or from an environmental watch group. These "real life" summaries bring meaning to textbook reading, and open up the world of science learning beyond the four walls of the classroom.

Text Complexity in Science

Along with today's current standards came the idea of providing increasing levels of text complexity. This idea is not entirely new; however, the Common Core English Language Arts Standards have brought it to the forefront of teachers' planning and implementation. Text complexity can be divided into three categories: *quantitative*, *qualitative*, and *reader and task*. Science teachers may already be familiar with quantitative measures. Perhaps the text resources in the classroom identify a guided reading level, a grade equivalency, or a Lexile® level. All of these measures attempt to assign a specific quantity to a text's level of difficulty. They might be based on a combination of word length, sentence length, and frequency of words. Science teachers can quickly discover the Flesch-Kincaid reading level of any text by following the directions in the following section. However, these quantitative measures do not take into account other factors that contribute to a student's success with a particular text. These are the reader and task variables and qualitative measures that balance the quantitative measures to determine a text's complexity.

To find out the Flesch-Kincaid reading level of any reading materials, teachers can use Microsoft Word®. Designed by Rudolph Flesch over 50 years ago, these scores can provide a quick way to assess the reading level of any text. The higher the Flesch Reading Ease score, the easier it is to understand the document. The Flesch-Kincaid Grade Level provides a grade level approximation of the text.

1. Be sure the readability option is active. Go into the **Preferences** or **Options** menu.

2. When the **Preferences/Options** window opens, click on the **Spelling and Grammar** tab.

3. Near the bottom of the Spelling and Grammar window, look for a box labeled **Show readability statistics**. Check the box and click the **OK** button.

4. Type at least two paragraphs of text into the Microsoft Word® program.

5. Choose the **Review** menu, and click on **Spelling and Grammar**.

6. When the program finishes running the spelling and grammar check, a window with the readability statistics will come up. These include the Flesch Reading Ease and Flesch-Kincaid grade level.

Reader and task variables include the reader's age and interest, the background knowledge he or she brings to a text, and the readiness of the reader to learn the content. Qualitative factors include levels of meaning, text structure, language conventions and clarity, and knowledge demands. A first-grade teacher would likely not subject his or her students to a nonfiction text about haploids and diploids. First, the readability level of the text is likely too high. Even if the readability level is within an acceptable range, say around the second-grade mark, the content is inappropriate. First graders are just learning about life cycles. They are not ready to learn about life at the molecular level. Likewise, an eighth-grade teacher would not want to provide his or her students a text written at the second-grade level explaining the basic differences between solids, liquids, and gases. By eighth grade, the science teacher can safely assume that his or her students have at least a general foundation of the concept of states of matter. Some students may need a review before embarking on the molecular structure of select gases. In this case, the teacher would want to use a text with a complexity best suited for his or her students.

Sometimes, the exact informational text teachers want to use with students is only available online. However, when teachers make these types of text selections, they may be written at a readability level that is too high for a particular group of students. For example, a third-grade teacher may find the perfect text resource online related to the sun and stars. However, the text as it stands is written at the middle-school level. In this instance the teacher can copy and paste the text content into a word-processing program, and modify it slightly from its original state to reduce the readability level. Simple steps, such as shortening sentence length or substituting simpler words, can significantly reduce the readability level without compromising the content.

Supporting Student Learning

As mentioned, informational text can be challenging to students. They may lack the background knowledge or experiences to bring meaning to technical language within the text. This might also cause students to struggle with content-specific vocabulary. Science teachers should remember that, in general, the reading levels in most science textbooks are usually one or two grade levels above the actual grade identified for the book (Barton and Jordan 2001, iii). Teachers can support students as they read to comprehend the informational text placed before them. This supports students' literacy learning while also enhancing their understanding of content.

Standards 5 and 7 of the Common Core Reading and Informational Standards require students to be able to analyze the structure of texts as well as the text features used to support texts. This includes the organization of the text itself (e.g., compare/contrast or cause/effect) and text features such as headings, subheadings, captions, diagrams, and illustrations. According to Anne Shemkovitz (2014), having students explicitly study the features of nonfiction text can help them transition into reading-to-learn using nonfiction text. Students should regularly use the table of contents, chapter headings, subheadings, illustrations, captions, diagrams, glossary, and/or index to efficiently locate information. Model how to read a section of a chapter. Tell students that when a text mentions a figure or graph, they should look at it and read the caption. Students may have little knowledge as to how an image supports the written text. Even if students actually *do* look at an illustration, they may not read the text within it or note any information in the caption. Discussing the purpose of special features within the text can help increase students' understanding of how a text is structured as well as the science concept under study.

> *A fifth-grade teacher wants to introduce a nonfiction trade book about plate tectonics to her students. One of the topics is "hot spots." The teacher begins by having students look through the index, glossary, and table of contents to find which pages students might find information about hot spots. Then, students turn to the identified pages and talk with a shoulder partner about the information presented in the features on the page, such as illustrations, captions, charts, headings and subheadings. Finally, students predict what types of information they will read about in this section, then read to confirm their predictions.*

Directed Reading/Thinking Activity (DRTA)

One reading strategy that keeps students focused on the text is the directed reading/thinking activity (DRTA). This is best conducted with a small group of students. DRTA activities "encourage students to be active and thoughtful readers and strengthen critical thinking skills" (Reading Rockets 2014, under "DRTA"). With DRTA, the teacher guides students to use text features to make a prediction about what the text might be about. This starts with the front cover and table of contents. Students use headings, subheadings, illustrations, charts, diagrams, and other text features to predict what they will read about. Students then read the text to confirm their predictions. The teacher redirects them by asking them to confirm, with justification, whether or not their initial ideas were correct. Then, the teacher guides students to make predictions about the next section or set of pages. The process continues until the end of the text. The teacher constantly and consistently sets the purpose for reading after each segment of text.

Use pre-reading strategies such as, brainstorming, class discussions, and concept and mind mapping to help students prepare to read.

The students use the information to confirm or correct their predictions, citing text evidence to support their conclusions. Once modeled successfully, the teacher can instruct students to use this strategy when engaging in text independently.

Previewing Text

Nonfiction text can be overwhelming due to the quantity of information, the challenging academic vocabulary, and vast breadth of content. Some texts may not be written in a very engaging manner or in student-friendly language. Students can come to a text with little or no background knowledge. This places them at a disadvantage because they do not have experiences upon which to anchor the information. One way teachers can help students organize the information in a chapter book is to provide a blank visual of the introductory page of each chapter (Ogle, Klemp, and McBride 2007). An example can be found in Figure 6.2. The blank template should show empty boxes or drawings of the main features that mirror the page. The page might include a sidebar that lists a key question or big idea for the chapter and important vocabulary. Dark bars represent the chapter title, headings, and/or subheadings. A long box represents a timeline and lines indicate where text would be placed. A blank box represents a picture with a caption beneath it. Students can work with partners to match the text supports with the blank template and label the template with words from the text. The class should then engage in a discussion regarding each text feature. This will help students understand the purpose of each feature and how it provides a meaningful organization of the text.

Figure 6.2 Previewing Texts

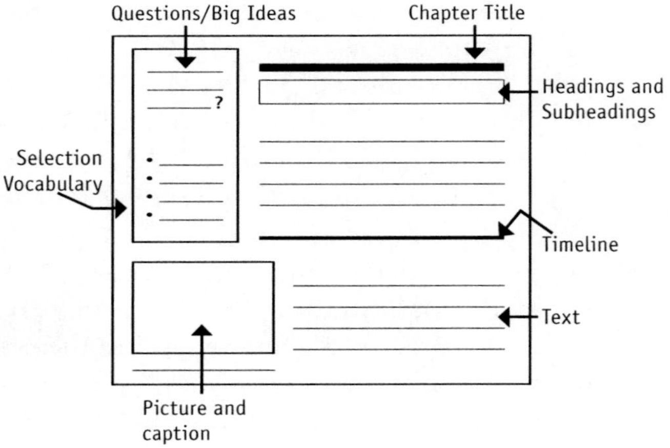

Paired Reading

In many classrooms, students commonly take turns reading aloud from their textbooks, one by one in a whole-class setting. When one student finishes a section, another begins reading. There are times when having one student read a passage aloud while others listen is appropriate. However, a teacher needs to ask: What are the other students doing while one is reading? Are all students focused on the passage? Are they just waiting for their turn to read? Are they looking at the book while thinking of something else? According to Patricia S. Koskinen and Irene H. Blum, students should "have an opportunity to read contextual materials a number of times so they can experience fluent reading" (2006, 225). Paired reading is a strategy in which all of the students are more involved and interactive. In this activity, students read aloud and talk about the reading with partners. Teachers should model their expectations for paired reading before setting partners to their work. First, Student *A* reads a short section of text aloud. Once completed, Student *B* summarizes what has been read. Student *B* can also ask questions of the reader. When the pairs have read and discussed their initial readings, they switch roles. Student *B* reads the next segment of text while Student *A* summarizes and asks questions. This way, *all* students in the class are actively engaged in the reading. Toward the end of the reading selection, the teacher brings the students together for a class discussion, and everyone can share what they have learned. As students become more proficient with the process, longer sections of text can be assigned.

In addition to paired reading, teachers might place students in groups of three. If a student is a struggling reader, he or she can always assume the role of active listener and still be responsible for asking questions and summarizing. In addition to reading textbooks, students can be paired to read different materials. They may read informational texts such as newspaper articles, informational articles, or primary sources. The materials will likely be on different reading levels to accommodate differences in the learning needs of students. If different groups read different texts, the teacher should be sure to allow students class time to share what they have learned. This way, everyone benefits from this new information, and everyone, regardless of ability level, has the chance to contribute to the class discussion.

Annotating Text

Textbooks and other expository texts include a great amount of information. So, how can teachers help students distinguish between what is important and what is supporting or interesting material? This is a very challenging skill for students at all levels, and teachers must constantly revisit this topic. Often, words such as *important*, *main*, and *first* give students clues about what is most important. However, these signal words are sometimes missing. Text marking is an interactive way of having students make connections, determine important information, look for information that is interesting, and ask questions if they do not understand.

A teacher begins by selecting an appropriate piece of text to display to students. The coding system can be found in Figure 6.3. As the teacher reads the passage aloud, "★" marks important information, "!" designates interesting information, and "?" is used when questions arise. When the teacher makes personal connections to the text or connections to material previously covered, the "+" symbol is used.

Figure 6.3 Coding for Text Marking

★ Important information

! Wow! Interesting information?

? I don't understand

+ This reminds me of...

Once the strategy is taught, students read short passages independently. They use the coding system on sticky notes to mark sentences or phrases in the passage. When students have finished the passage, they are organized into small groups to discuss what they found important, interesting, confusing, or connected to what they already knew. They discuss similarities and differences in their coding and share justifications for their markings. Finally, students are brought together as an entire group to compare what the groups found important or interesting. As the passage is discussed, the students should defend their ideas about what is most important and what is simply interesting information. The teacher should also list questions and clarify any misunderstandings. The discussion helps the teacher informally assess learning and guides further instruction.

Making Inferences

In science, textbooks and other examples of expository material require students to use skills of analysis to make inferences. "Inferential thinking occurs when text clues merge with the reader's prior knowledge and questions to point toward a conclusion about the underlying theme or idea in a text" (Harvey and Goudvis 2000, 23). In other words, students use what they know to fill in the gaps when they read informational text. As the texts become more difficult, the expectation for them to make inferences about what they read can become a very challenging task.

Even the youngest children make inferences in their everyday lives. One way teachers can help with challenging informational reading is to remind students of these primary-level experiences and show them the parallels in making inferences as they read text. Often, elementary-school teachers begin by using picture books and asking students to use the words in the text and the visuals to make inferences. To teach this concept at all age levels, teachers must first find out what students already know and then connect that information with new information in the text. When teachers come upon statements in text that lend themselves to making inferences, they should model this by thinking aloud to show their thought processes. Students should identify word clues that help make accurate inferences as well as discuss words that contribute to their inferences. In science, teachers can provide experiences where students must use their observational skills to draw inferences. The teacher might ask "What happens when something touches a worm?" Students can make predictions, then try it on their own. The teacher can record the students' observations on the left side of a T-chart (e.g., it moves away, it wiggles). If students make an inference (e.g., it doesn't like being touched; the light scares it), these would be recorded on the right side of the T-chart. Now, the teacher can lead a discussion about why some statements are observations (they use the senses), while others are inferences (they are based on more than just observations). Now, when students read facts about worms, they can connect their observational facts to their inferential thinking. Perhaps the text will confirm that indeed, worms instinctively move away from outside stimuli, which the students would have observed. The text might even justify why this happens, and students can relate these ideas back to the inferences they made based on the activity.

An Inferencing Strategy: "It Says... I Say... and So..."

It Says . . . I Say . . . and So . . . is an inferencing strategy developed by Kylene Beers in 2003. This graphic organizer is used to teach the skill of making inferences. As always, the teacher must model the strategy and revisit it often, thereby providing students with ample opportunities for practice. Students first begin with a question about the text that requires some inferential thinking. This step is the key to the process. Initially, teachers can read the passage orally as the students follow along. Then, the class identifies text-based information related to the question. They record this in the "It Says" column. Students should use words directly from the text. This will help them meet the expectations set forth by the Common Core English Language Arts standards. Next, students verbalize their own background knowledge and experiences. They place this information in the "I Say" column. The inference "and So" is made when the students combine text-based information from the "It Says" column with what they know from the "I Say" column. Figure 6.4 has been completed showing a question that causes students to make an inference about the lives of volcano chasers.

Figure 6.4 Sample It Says... I Say... and So...

Question	It Says . . .	I Say . . .	and So . . .
Read the question.	Find information from the text to help you answer the question.	Consider what you know about the information.	Put the information together from the text with what you know to answer the question.
What can we infer about the lives of volcano chasers?	They travel the world studying volcanoes. They measure earthquakes. They check for dangerous gases spewing from volcanoes.	They use what they find to warn people of the dangers in the area where they live.	People who are volcano chasers have a very important job and they end up saving many lives.

Summarizing

Teachers often say that summarizing is one of the most difficult skills for students to learn (Duke and Pearson 2002). However, summarizing is a valuable learning skill. When students can summarize, they recognize the main ideas and can eliminate unimportant information. They are required to think deeply about the information they read and use skills of analysis. They refine their vocabularies, and when they put the main ideas into their own words, they also better remember the content. Summarizing activities can be as simple as having students write a one-sentence summary to explain a concept, process, or skill. Summaries can be as complex as having students synthesize information from a multitude of resources into one coherent, focused essay or proposal. Teachers should model how to summarize the key points from a reading passage, as shown in Figure 6.5. Textbook passages and articles likely hold clues to the main ideas in the headings and subheadings.

Figure 6.5 Sample Graphic Organizer for the Topic of Force and Motion (grade 2)

Pushes	Pulls
Someone pushes me on the swing	I pull a wagon
I push open a door	I pull up flowers
I push the keys on the computer	Gravity pulls me down a slide

There are two kinds of forces. They are a push or a pull. Someone might push me on a swing. I can push open a door. I push keys on a computer. I pull a wagon. I pull up flowers from the ground. Gravity pulls me down the slide. I use pushes and pulls every day.

Walter Pauk (2007) developed a note-taking system in the 1950s when he was an education professor at Cornell University. His system of note taking, now referred to as Cornell notes, is a popular manner in which students can record key words and/or questions, important ideas, and a summary of the information on one sheet of paper. Teachers can also have students complete Cornell notes to help them learn to identify main points and supportive details, as shown in Figure 6.6. They can also have students complete a graphic organizer to help them organize (and subsequently summarize) information as shown in Figure 6.7.

Figure 6.6 Cornell Notes Sample Outline

Course	Name	Date
Key Points	**Notes About Key Points**	
Summary		

Figure 6.7 Sample Concept Map Summary on the Topic of Plate Tectonics (grade 7)

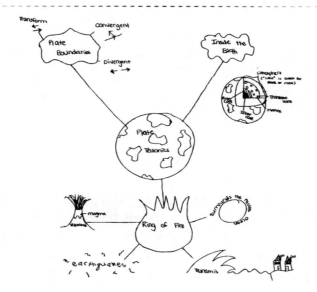

Jot the Gist

In this strategy, students explain the "gist" of a piece of text by summarizing it in a few words (Moore et al. 2006). They look for important information, eliminate unimportant information, remove redundancies, and write brief summaries in twenty words or less. One way to introduce this strategy is by using a short newspaper article with approximately three paragraphs. To help students look for the most important information, tell them to begin by focusing on who, what, when, where, why, and how of the information read. After students have read the article, conduct a class discussion. Students recall the most important information in the text and record this information. Then, have students work as a class to take the important words and condense them into a clear summary of about 20 words. Students will need to think of synonyms and other words that incorporate the meaning of one or two of the words in their initial list. As students become more familiar with the "gist" strategy, they can work in pairs and then, ultimately, independently. Students also enjoy giving newspaper articles new headlines. This requires them to use the summarized information and condense it even more succinctly.

Magnet Summaries

Magnet summaries are another way to teach summarization (Buehl 2001). Again, students look for the most important information in text, eliminate unimportant information, and then write summary sentences. The magnet analogy engages student thinking. Just as magnets attract metal objects, magnet words attract key information. This strategy works well when using a textbook. Ask students to read a short section of the text on a given topic and look for key words that explain the topic. Lead the students in a class discussion and write the important information about that topic around the word. Discuss the information that students want included on the magnet. Then, model how to write a brief summary sentence using the important words. Students should understand that they might not use all the words because some are more important than others. The summary statement should be approximately twenty words in length. As students become more comfortable with this strategy, they may work in small groups or pairs to write magnet summaries for

Sample Summary Sentence

Tornadoes are fast moving funnels touching the ground (and sometimes water), produced by thunderstorms and rated on a Fujita Tornado Scale.

other parts of the text. Students can also put their magnet summaries on index cards. The information is on one side of the card, and the sentence summary can be written on the back of the card. When a number of cards are created, they become an excellent study guide for a larger topic.

Direct Vocabulary Instruction in Science

Vocabulary is integrally linked with background knowledge. Once students have built some background knowledge, teachers must move on to teaching words and concepts explicitly. "Systematic vocabulary instruction is one of the most important instructional interventions that teachers can use, particularly with low-achieving students" (Marzano, Pickering, and Pollock 2001). In addition to building background knowledge, explicit vocabulary instruction increases reading comprehension, helps students communicate more effectively, improves the range and specificity of student writing, enables students to communicate more effectively, and helps students develop a deeper understanding of concepts (Allen 1999).

The science discipline has a unique vocabulary, and therefore, teachers must build vocabulary instruction into their planning. "Teaching words well entails helping students make connections between their prior knowledge and the vocabulary to be encountered in the text and providing them with multiple opportunities to clarify and extend their knowledge of words and concepts during the course of study" (Vacca and Vacca 1999, 319). With so much vocabulary in science, where do teachers begin? It is important to focus on specific words that are related to what students will be learning. While teachers cannot teach all the content words students need to know, they must strategically pick a few that are essential for understanding major concepts in a unit.

Looking up words and writing their definitions does not help students learn vocabulary. Instead, teachers must provide a variety of opportunities for students to interact with words. Janet Allen (1999) advises that words be used in a *meaningful context* between 10 and 15 times. More recent research also advises that students create pictures and other graphic representations of words, be able to compare and contrast words, classify them, and create metaphors and analogies (Marzano, Pickering, and Pollock 2001). The following are some vocabulary instructional strategies that can help science teachers incorporate direct vocabulary instruction into their days.

Word Walls

Word walls can be seen in classrooms from the primary level to high schools. For the purposes of this book, the focus will be the use of content word walls that help develop academic vocabulary. Some word walls are arranged in alphabetical order while others are arranged by topic. However, it is not enough to just have words posted in the classroom. Rather, word walls must be interactive. Key vocabulary words must be carefully chosen and introduced, then gradually added to the walls (Allen 1999). The words must be posted and visible throughout a given unit, where students can refer to them often to use them in their discussions and writings. The words should also be displayed on cards, so students can easily manipulate them and make connections among the words. For instance, students could:

- sort and classify them in various ways;
- regroup words when they look for causes and effects;
- look for ways to compare and contrast words;
- find synonyms and/or antonyms;
- examine positive and negative connotations;
- use them in journal entries;
- create picture dictionaries;
- identify the "most important" words and give their reasons; and
- explain a word's importance to a unit of study.

When words are used in warm-up exercises or at the end of lessons, teachers also have an excellent informal assessment tool.

Word Questioning

Word Questioning asks students to apply higher levels of thinking to figure out unknown words. It employs Bloom's Taxonomy to help students analyze, comprehend, apply, synthesize, demonstrate knowledge, and evaluate what they know about a particular concept (Bloom and Krathwohl 1956; Anderson and Krathwohl 2001). Consider a unit on electricity. Figure 6.8 shows the graphic organizer for the word *electromagnetism*. Students complete the surrounding boxes to (1) see what parts of the word they might recognize; (2) make a prediction about the meaning of the word; (3) give examples of what it is and what it is not; (4) evaluate its importance; (5) make connections to what they already know; and (6) think about when, where, and under what circumstances they would find the word.

Figure 6.8 Sample Electromagnetism Word Questioning

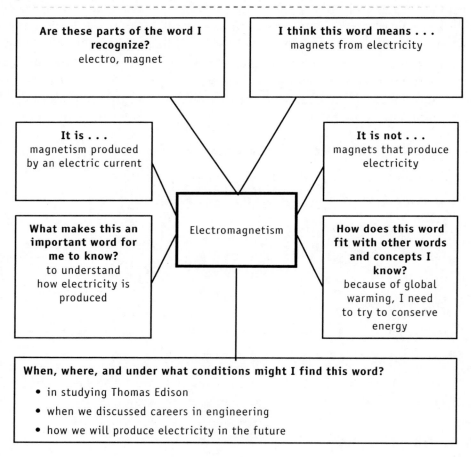

Comics

A creative way to have students use and practice science-related vocabulary is to have them summarize a process (such as the water cycle) or a concept (such as gravity) using comics (Kopp 2010). As part of classroom instructional time, or as homework, students can create original comics using key characters or simple pictures with speech bubbles to appropriately use content vocabulary in context. If they complete the activity on note cards, students can use their comic content to study their words independently or they can practice them with other students when they share their comics in a paired setting.

Frayer Model

The Frayer Model (Frayer, Frederick, and Klausmeier 1969) is a strategy that helps students understand concepts. It is an organized graphic that can be used as a basis for writing, even with younger students. It gives students opportunities to define and describe a word or idea. Students also demonstrate their understanding by providing examples and non-examples. Figure 6.9 shows an example using the concept of *atomic arrangement*. Students define the topic in their own words, list essential characteristics, and then add examples and non-examples. In variations of this model, students replace the information in one of these boxes with other tasks, such as including an illustration or using the word in an original sentence.

Figure 6.9 Sample Atomic Arrangement Frayer Model

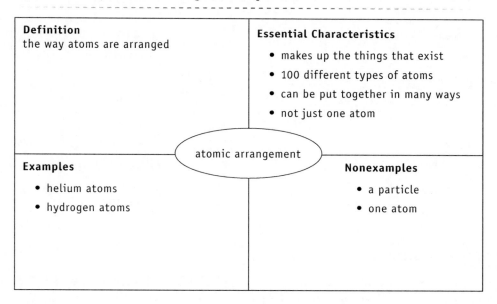

Making Comparisons

Because students must learn many terms and concepts, ones they tend to only see in science class, an effective vocabulary activity is one that has students compare and contrast information. Research confirms that one of the most effective ways for students to retain content information is to have them make comparisons between ideas (Marzano, Pickering, and Pollock 2001). Often, teachers use a Venn diagram, but other graphic organizers can be just as effective. When students study landforms, they must examine forces that change Earth's surface. The H–Diagram in Figure 6.10 shows a comparison between constructive and destructive forces. As students examine the critical information about the two forces, they list the differences on the two sides of the *H* and the similarities in the crossbar. Ogle, Klemp, and McBride (2007) also suggest using a Y–chart to illustrate differences and similarities between terms. Students list differences between two concepts or terms in the top part of the *Y* and the similarities in the base of the *Y*.

Figure 6.10 Sample Forces H-Diagram for Comparisons

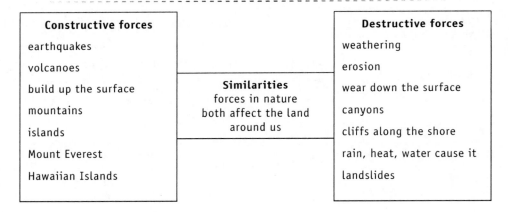

Concept Circles

Another technique for assessment is the use of concept circles (Vacca and Vacca 1999). Figure 6.11 shows a circle divided into four parts, but it can be divided into six, or even eight, parts depending on the concept. With concept circles, students must know the meaning of the words in the sections, analyze the relationships among the words, and think of the concept that ties the words together. In this example, the concept is *a galaxy*.

Students understand that a galaxy called the Milky Way consists of planets, stars, and moons.

Figure 6.11 Sample Galaxy Concept Circle

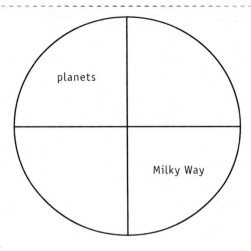

Teachers can use other variations of concept circles. A teacher can give students a circle that has words in three sections and one blank section. Again, students must know the meanings of the three words, understand their relationships to one another, and identify the concept. Then, they must add a fourth word in the blank section that also relates in a similar way to the concept. In another variation, a teacher can give students blank organizers but tell them the concept. Students then need to fill in the sections of their circles with words that describe the concept. They then must justify how the words are related to the concept.

Using Words in Context

It is nearly impossible for teachers to analyze every new word that students encounter or even teach all the words that students do not know. So how can teachers help children become independent learners and figure out new words from context? Often when students encounter new words, they just skip them. Unfortunately, that strategy does not serve learners well. Teachers must give students the strategies for figuring out new words in context. As the teacher reads a passage orally or as students tackle difficult text, they should be aware of the strategies they are using to figure out new words and be able to apply those strategies. Teachers should model how to understand new words in context.

- Use background knowledge and experiences as a guide.

- Use signal words such as *for example* or *including*. At other times, the author will restate the word, using phrases such as *in other words* or *also called*. Words like *consequently* or *because* signal a cause–and–effect relationship. Comparisons may be signaled by *like* or *similar to*, and contrasts may be signaled by *but*, *however*, and *in contrast*.

- Teach students about common prefixes, suffixes, and root words. These can help students determine the meanings of unfamiliar words.

- Use surrounding text features. Graphs, titles, pictures, and footnotes all help students find the meanings of new vocabulary words.

Mine, Yours, Ours

The Mine, Yours, Ours technique helps students explore key concept words in context. This strategy may be used to introduce big ideas, such as renewable and nonrenewable resources, light, ecosystems, force and motion, etc. The teacher provides students with the word, along with one or two sentences using the word in context. The sentence must provide adequate context clues for students to infer its meaning. Students independently write what they think the word means in the "Mine" column. Then, they turn to a partner and share their definitions. They write the other student's definition in the "Yours" column. Then, the class discusses the students' ideas to come up with a class definition of the word, which students write in the "Ours" column. (In another version of this strategy, the teacher and students can record an official dictionary definition. They would modify the final column to be a "Theirs" column.) Finally, students collaborate to compare the definitions, and summarize their thoughts at the bottom of the page. Figure 6.12 demonstrates what this might look like in a slide show presentation.

Figure 6.12 Sample Mine, Yours, Theirs

Mine, Yours, Theirs		
Plate Tectonics		
The multi-million dollar seismic equipment was helpful. But the geologist relied more on her understanding of plate tectonics to help predict the next major earthquake.		
My Definition:	**My Neighbor's Definition:**	**Glossary Definition:**
Something that helps scientists predict earthquakes	Parts of the earth, like rocks and dust	A scientific theory. It explains the way that land separated into today's continents from Pangaea. It is also the study of land movement, which causes earthquakes.
How These Definitions Compare:		
I really had no idea what plate tectonics were. My neighbor and I figured out that it had something to do with the earth and earthquakes. It really is about how the earth moves.		

Writing Standards in Science

Writing promotes learning through the use of reflection and discussion. Writing Across the Curriculum states "Only by practicing the thinking and writing conventions of an academic discipline will students begin to communicate effectively within that discipline" (Michigan Science Teachers Association 2014, 3). Science teachers may consider themselves to be in a prime position to help students see the value of writing for a specific purpose and to help them perfect this craft. The Common Core State Standards Initiative also supports writing within the science classroom by stating "To build a foundation for college and career readiness, students need to learn to use writing as a way of offering and supporting opinions, demonstrating understanding of the subjects they are studying, and conveying real and imagined experiences and events" (CCSSO 2010, 41). The following are some examples of options available to utilize writing as a means to learn.

Interactive Notebooks

First and foremost, students may respond to learning by summarizing or recording information in a science journal. The purpose behind an interactive notebook is to help students organize and synthesize information given to them in class. This strategy also helps students with different learning styles comprehend the information being presented because it allows them to put what they have learned into their own words. Caroline Wist of The College of William and Mary states "Interactive notebooks can help students process information, study and review for assessments and personalize the content being presented" (2014, 7). The science journal is a safe place for students to jot notes, thoughts, ideas, information, or other relevant matter as it relates to their own personal learning. The notebook should not be used for students to "copy" notes that the teacher records. This does not require students to think on their own, nor does it lead to meaningful learning. The use of a science journal helps students to organize their thoughts and ideas which can lead to increased comprehension of the concept by allowing them to reflect on their writing.

Graphic Organizers

Another option to have students write in science is to provide them with outlines or other graphic organizers. For students who are English language learners or those who require differentiation, completed outlines and organizers provide additional support for success. Science students should have a study guide handy any time they engage in an activity such as when they watch a video, listen to a guest speaker, or complete a lab or investigation, whether it be at the basic confirmation level or at the most complex open level. This holds students accountable for their learning at all times. These summaries also make great review pages when students need to study for an end-of-unit assessment. Figure 6.13 shows an example of a flow chart organizer that can be used for sequencing activities. Figure 6.14 can be used to show the relationship between science terms or concepts. Figure 6.15 is an example of a T-chart organizer, which is a helpful tool to jot down observations when gathering data during an experiment. Lastly, Figure 6.16 shows a sample Scientific Method organizer which is another tool for recording information during a science experiment.

Figure 6.13 Sample Flow Chart Organizer

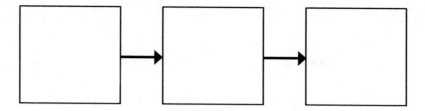

Figure 6.14 Sample Web Organizer

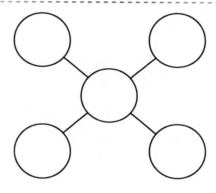

Figure 6.15 Sample T-Chart Organizer

Trial	Observation
1	
2	
3	

Figure 6.16 Sample Scientific Method Organizer

Problem	
Background Research	
Hypothesis	
Experiment	
Results	
Conclusion	

Essays, Lab Reports, and Creative Projects

Finally, students can complete formal essays, independently completed lab reports and lab summaries, and creative projects to demonstrate their understanding of science concepts in written form. Lab reports generally contain a title, abstract, an introduction which states a hypothesis, materials and methods used to test the hypothesis, results of the experiment where data is provided, a discussion section where students talk about whether or not the experiment proved or disproved their hypothesis, and a literature cited section.

Another way to have students write within the science classroom is through the completion of creative projects. For example, students learning about ecosystems may design and create a model of an ecosystem, which includes biotic and abiotic factors, producers, consumers, decomposers, and an energy source. To determine what students really know about ecosystems, they can write an interview between two animals which references each part of the ecosystem. They could also write a "Live-on-the-Scene" news report about something that is happening with this ecosystem that references each part. With a well-designed rubric or grading scale, the teacher can use this written project as a summative assessment (an example project outline and evaluation criteria may be found in Appendix D).

> Consider using digital media when assigning writing projects to students. Classroom blogs, class wikis, and class websites are kid-friendly means for students to publish factual summaries, photos and illustrations, multi-media presentations, graphic organizers, and notes.

Regardless of the writing assignment science teachers require of their students, they will, along with their language arts counterparts, hold students accountable for the Common Core English Language Arts Writing Standards. A general summary of these standards may be found in Figure 6.17.

Figure 6.17 Common Core College and Career-Ready Anchor Standards for Writing

CCRA.W.1	Write arguments to support claims in an analysis of substantive topics or texts, using valid reasoning and relevant and sufficient evidence.
CCRA.W.2	Write informative/explanatory texts to examine and convey complex ideas and information clearly and accurately through the effective selection, organization, and analysis of content.
CCRA.W.3	Write narratives to develop real or imagined experiences or events using effective technique, well-chosen details, and well-structured event sequences.
CCRA.W.4	Produce clear and coherent writing in which the development, organization, and style are appropriate to task, purpose, and audience.
CCRA.W.5	Develop and strengthen writing as needed by planning, revising, editing, rewriting or trying a new approach.
CCRA.W.6	Use technology, including the Internet, to produce and publish writing and to interact and collaborate with others.
CCRA.W.7	Conduct short as well as sustained research projects based on focused questions, demonstrating understanding of the subject under investigation.
CCRA.W.8	Gather relevant information from multiple print and digital sources, assess the credibility and accuracy of each source, and integrate the information while avoiding plagiarism.
CCRA.W.9	Draw evidence from literary or informational texts to support analysis, reflections, and research.
CCRA.W.10	Write routinely over extended time frames (time for research, reflection, and revision) and shorter time frames (a single sitting or a day or two) for a range of tasks, purposes, and audiences.

—CCSSO (2010, 18)

Conclusion

Even in an inquiry-based science classroom, students can and should be required to read and write in response to learning. Teachers have many options when it comes to using the most appropriate nonfiction text resources with students. Although some text resources could prove challenging to some students, teachers can use differentiation strategies such as cooperative learning, direct and guided reading instruction, direct vocabulary instruction, and graphic organizers to support student learning in science. The added element of writing in response to learning allows students to record thoughts, ideas, and questions. It encourages independent thinking and creative thought, and students can better find value in the work they complete as part of their journey to becoming scientifically literate.

Stop and Reflect

1. How do today's standards for writing positively impact science instruction?

2. What are the most prevalent reading materials that students use in your classroom? What materials might you try to use more often?

3. What types of writing tasks can you assign students as part of their science learning that support the current standards?

4. What differentiation strategy or strategies (for reading, writing, or both) seem the most realistic given your particular teaching situation or setting?

Chapter 7

The Math and Science Connection

 "Go down deep enough into anything and you will find mathematics."

—Dean Schlicter

Every year, elementary school teachers open their math books and begin their unit related to measurement. Each year, elementary school students are expected to master how to measure correctly with the correct instrument, including linear measures (inch, centimeter, meter, etc.), volume (cup, liter, kiloliter), and time (hour/half-hour, minutes, seconds, elapsed time). In later years, teachers include area (square units) and volume (cubic units). Every year, the subsequent grade levels' teachers wonder, "Did last year's teacher even teach this concept? These students are still struggling with this concept." Likely, last year's teacher did in fact teach the concepts of measurement and time. However, how many opportunities do students have to actually measure and tell time on a regular basis? If they have never baked at home and all they have are digital clocks, the answer is, very few. So these ideas of measurement and time, although perhaps mastered at one point during the previous school year, may not have been practiced or reinforced; these concepts have not been transferred to students' long-term memory. Fortunately, science concepts and inquiries use an abundance of mathematical concepts, including operations and algebraic thinking, number and operations in base ten, measurement and data, and geometry. Coincidentally, all these concepts are part of the Common Core State Standards for Mathematics (CCSSM). At the elementary level, the standards call these big ideas "domains." This chapter explores the natural marriage of math and science, and demonstrates how teachers can provide learning opportunities in science to strengthen and support math skills.

Standards for Mathematics in Science

Standards for mathematics define what students should know and be able to do with regard to mathematics skills. Standards are generally organized by major concepts. For example, in middle school, concepts may include ratios and proportional relationships, the number system, expressions and equations, functions, geometry, and statistics and probability. In high school, the standards could be more course-oriented. For example, the Common Core State Standards for Mathematics include reference to eight mathematical practices. These practices closely mirror the science and engineering practices that are inherent in the Next Generation Science Standards (NGSS).

Standards for Mathematical Practices

- Make sense of problems and persevere in solving them.
- Reason abstractly and quantitatively.
- Construct viable arguments and critique the reasoning of others.
- Model with mathematics.
- Use appropriate tools strategically.
- Attend to precision.
- Look for and make use of structure.
- Look for and express regularity in repeated reasoning.

—CCSSO (2010, 6–8)

As students work in an inquiry-based science classroom, they inevitably will apply all of these mathematical practices. However, as stated in Appendix L of the Next Generation Science Standards "The three CCSSM practice standards most directly relevant to science are practices 2, 4, and 5" (2013d, 3). Additionally, students will use computation skills, measurement, and data analysis as a natural part of an inquiry lesson. It bears mentioning here, science teachers should be careful when organizing science inquiry activities that may require students to use math skills that are not part of their previous learning. For example, a science teacher may have students construct a model of a building or structure to test its strength against hurricane- or tornado-force winds. This type of activity motivates students, since they are challenged to "beat Mother Nature." They strive to create something that will remain standing after having been subjected to high winds (in this case, a gas blower). This activity, to be completed to scale, involves students' understanding of ratios and proportions. This is an advanced math skill that does not appear in students' curriculum until middle school. So elementary teachers may need to reconsider the need to build the structure to scale, or they can take time to teach students about ratios and practice using them. In another example, second-grade students may use digital scales that measure to tenths or hundredths of grams. However, the concept of decimals does not appear in the Common Core State Standards for Mathematics until 5th grade. In this instance, the teacher may choose to measure in milligrams, or guide students more directly as they encounter these numbers so that they understand and make sense of the data they collect. The Next Generation Science Standards developers worked with Common Core State Standards for Mathematics developers to ensure a proper grade-by-grade alignment between the concepts students learn in both science and math. The table shown in Figure 7.1 from Appendix L of the Next Generation Science Standards (2013d) illustrates key math topics relevant in science and the grade levels where these math concepts are first introduced.

Figure 7.1 Key Math Topics Relevant to Science as They Appear in CCSSM Standards

Number and Operations	Grade First Expected
Multiplication and division of whole numbers	3
Concept of a fraction a/b	3
Beginning fraction arithmetic	4
The coordinate plane	5
Ratios, rates (e.g., speed), proportional relationships	6
Sample percent problems	6
Rational number system/signed numbers—concepts	6
Rational number system/signed numbers—arithmetic	7
Measurement	**Grade First Expected**
Standard length (inch, centimeter, etc.)	2
Area	3
Convert from a larger unit to a smaller in the same system	4
Convert units within a given measurement system	5
Volume	5
Convert units across measurement systems (e.g., inches to cm)	6

—NGSS (2013, Appendix L)

Relevance of Math in Science

In math classes, students must understand the relevance of the skill if they are to engage and learn it. Robert Marzano and Debra Pickering (2010) address this idea as it relates to student engagement. Students, when presented with any material, will ask themselves "Is this important?" If the answer is no, they will be less engaged. Marzano and Pickering identify three major areas of focus when demonstrating instructional relevance to students. First, the content must be connected to the students' lives. Secondly, it must be connected to students' life ambitions. Thirdly, students must be provided opportunities to apply the skill and knowledge in varying contexts. This third idea, especially, is where science teachers can bridge the connection for students in the use of their math skills. As Ben Johnson reminds us, "When a teacher gives students a real science problem to solve—one that requires math tools—the teacher is giving the students a reason to use math. Math, then, becomes something useful, not something to be dreaded (Johnson 2011, under "Teacher Leadership").

The application of math skills in science may take many forms across the grade levels. Early learners may use counting or simple computation during an inquiry lesson. For example, a kindergarten standard requires students to describe measurable attributes of objects and compare objects by their measurable attributes (CCSSM: K.MD.A.1, K.MD.A.2). So, even as early as kindergarten, students begin to learn and apply measurement skills in science class. Students might measure and compare the lengths of flowers surrounding the building. They could even compare how high they can jump. An inquiry activity students could complete is planting flower seeds in class. The students can fertilize one flower and not fertilize another. As the flowers grow, they measure and compare the rate of growth of both. A sample inquiry question may be *How does fertilizer affect the rate of plant growth?* In first or second grade, students can make string phones with strings of different lengths. A sample inquiry question could be *How does the length of the string in a string phone affect the sound quality?*

Companies and organizations support science teachers as they integrate math into science. PUMAS (Practical Uses of Math and Science) is an online journal of math and science examples. Students in grades K–12 follow fictional, but true-to-life examples of situations that require creative problem solving of a scientific concept through the use of mathematics.

Middle- and upper-elementary students apply fraction concepts or more advanced (multi-step) computation skills. For example, as early as third grade, students must "recognize and generate equivalent fractions" (CCSSO 2010, 24). Students can apply this concept in science by identifying and comparing the life spans, in fractions of years, of organisms. During Science, Technology, Engineering, and Math (STEM) activities, teachers could assign cost values to the items needed to build a prototype. Then, students must calculate the cost-effectiveness of their design. When they work their revisions, can they make their design more cost-effective?

Middle- and high-school students may apply scientific formulas in chemistry and physics, and use algebraic concepts and measurement error as they recognize patterns and relationships among variables. In science, math process skills (what the Common Core refers to as mathematical practices) are also used for a myriad of purposes. Students should be able to reason abstractly and quantitatively to analyze patterns in nature. (Coincidentally, reasoning skills are both a Common Core State Standard for Mathematics mathematical practice and a science and engineering practice.) Students will likely use math skills (e.g., counting, graphing, comparing, applying measurement error) to accomplish this mathematical practice. In eighth grade, students must "construct and interpret scatter plots" (CCSSO 2010, 56).

Nature provides many opportunities for students to compare two variables to determine their relationship. For example, students might investigate how arm span compares to height; how the temperature of water compares to the mass of ice added; how velocity affects the stopping distance of an object; or how adjusting the temperature of a home affects the monthly electric cost. Figure 7.2 shows a scatter plot students might create as they compare the beginning height of a dropped rubber ball, and its resulting single bounce height. With this data, students would attempt to determine if a relationship (or correlation) exists between the height that a rubber ball is dropped with its resulting bounce height, and justify their findings. Figure 7.3 shows how students may have recorded and calculated the change in temperature of sand as it was moved to direct sunlight, then back into a shaded area. Once calculated, students would next determine whether the sand gained or lost heat energy more quickly, and justify how they know.

Figure 7.2 Sample Science-Related Scatter Plot, with Trend Line

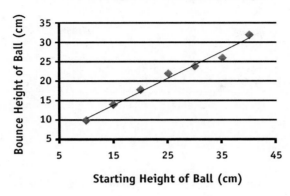

Ball Height vs. Bounce Height

Figure 7.3 Sample Science-Related Calculations

Temperature Change of Sand	
Starting Temperature	23°C
In Direct Sunlight	
2 minutes	25°C
4 minutes	27°C
6 minutes	30°C
8 minutes	34°C
10 minutes	36°C
TOTAL CHANGE (+/−)	+13°C
In Shade	
2 minutes	34°C
4 minutes	31°C
6 minutes	29°C
8 minutes	27°C
10 minutes	26°C
TOTAL CHANGE (+/−)	−10°C

Figure 7.4 further illustrates the relevance of math skills as students reason abstractly and quantitatively in science class to complete an inquiry lesson. In each instance, these activities require students to use mathematical skills and to apply most or all of the mathematical practices.

Figure 7.4 Examples of Mathematics Skills Used in Science Activities

Science Activity	Mathematics Skills Used
Grade 2: Do all daisies have the same number of petals? Pairs of students count the number of petals on a daisy. The class collaborates to discover how the daisies compare.	Counting and ordering numbers; looking for patterns in numbers
Grade 7: Can sunscreen protect paper from the sun's ultraviolet light? Students use colored construction paper, sunscreen, and sunlight to determine whether sunscreen can protect paper like it protects human skin when exposed to the sun's ultraviolet rays.	Using a scale and/or graduated cylinder to measure equal amounts of sunscreen; developing a rating scale to quantify the color quality of the paper; using a stopwatch to calculate the time the sun takes to diminish the color quality of the paper; creating a graph to compare the elapsed time in the sun to the color quality of each paper
High School Physics: How strong is gravity where I live? Students create a pendulum, and measure its period. Then, they use the formula for a period to calculate the force of gravity where they live.	Using algebraic equations to solve for an unknown variable; applying systematic error analysis when interpreting data

Reinforce Vocabulary

Another advantage for students whose science teachers recognize and demonstrate mathematical skills and practices is the fact that they will also reinforce math (and science) vocabulary. For example, in the second grade illustrated in Figure 7.4, students can reinforce the following math terms: *pattern, odd, even, digits, compare, less than, greater than,* and *difference.* This activity would most likely be linked to a unit on heredity and reproduction, including the life cycle of a plant, and the idea that offspring resemble their parents. Additionally, students can use the greater than and less than symbols (>, <) to compare the quantities among the flowers. Science terms that might be reinforced during this activity include: *characteristic, petal, offspring,* and *germinate.* In the middle school example, students can reinforce these math terms: *volume, surface area, congruent,* and *scatter plot* (or other graph). Science vocabulary would likely be related to the topic of the electromagnetic spectrum and light energy. In high school, students can reinforce these math terms: *radical, equation, variable,* and *measurement error.* During this investigation of pendulums, students can reinforce their understanding of the terms and equations for *force, motion, acceleration, period,* and *frequency.*

Why should math teachers have all the fun with digital manipulatives? According to Michelle Davis, writer for *Education Week*, "More than ever, teachers of mathematics and science say, when digital tools are incorporated into the curriculum, the change motivates students to get through the drudgery and uncertainty of data collection to the payoff of results that simulate and showcase the theories they're studying. Plus, more sophisticated use of technology in math and science classes offers students opportunities to do more hands-on work and experience what professionals are actually doing in laboratories" (Davis 2007, under "Digital Directions"). Science teachers can access online math tools as students make calculations and generate graphs in class. An online graphing calculator is available at Cool Math®. The National Library of Virtual Manipulatives has digital math tools for all functions: number and operations, algebra, geometry, measurement, and data analysis and probability.

Supporting Student Learning

The science teacher's job is not to teach students how to do math. However, if a particular inquiry activity requires students to use math skills that they have not yet mastered, have yet to be introduced, or that they have forgotten, the science teacher may need to conduct a mini-lesson on how to "do the math." This may include complex computations, place value to tenths or hundredths, deciding on the best organizational method to illustrate the data, or even simple measuring with various tools. A simple way to gauge whether students have mastered a particular math skill or not, such as subtracting across zeros, is to give a few sample problems, and have students then solve the problems independently using individual white boards. Once students turn the whiteboard toward the teacher, he or she may quickly scan the students' work and determine whether or not the skill is understood. If so, the teacher may use this information to strategically organize lab groups. If not, the teacher may spend class time teaching this skill, using math class to teach this skill, or allowing students to use calculators for the computations during the inquiry.

If mini-lessons are not a viable option, teachers can use cooperative groups as a means to allow access to advanced math skills for all students. For example, if twenty out of twenty-five students, based on a quick, formative assessment, know how to read a thermometer, teachers can organize lab groups so that the five students who lack this skill work among other students who have mastered this skill. The expectation, then, during lab time, would be that all students use the thermometer to measure temperature at least once. This way, all students have a turn, and the students who may struggle with this skill can receive a one-on-one tutorial from their lab partners.

Another way to use cooperative groups to support math skills in science class is to have each group collect the data collaboratively and individually complete the computations. Then, the groups collaborate again to compare their individual answers. Students would then discuss and justify the correct answer using mathematical language and processes if students in the group arrived at several different answers. A similar collaboration strategy is "think-pair-share." In this strategy, students think (or in this case, compute) to themselves, then turn to a partner to share, discuss, clarify, and justify their results.

One other strategy for supporting student learning in science is for the teacher to pull a small group of students to one area of the classroom. The teacher then becomes the facilitator of the group, using science learning as an opportunity to support math learning in a small-group setting. While the teacher is working with this small group of students, the other students are actively engaged in the work they must do to complete the inquiry. Having set clear expectations regarding proper behavior during inquiry instruction, the students not working under the teacher's direct supervision still know and abide by the expectations.

The idea here is that teachers do not want students to fail to complete an inquiry activity because they struggle with the appropriate math skill(s) it requires. By thinking ahead and knowing students' math abilities before beginning an inquiry activity, science teachers can ensure successful completion of the data element for all students, regardless of their math skills.

Ideas for Differentiating Math Instruction for Science Learning

- Conduct mini-lessons to teach the needed math skill(s) before beginning the inquiry activity

- Collaborative learning: strategic grouping; individual practice/collaborative comparison; think–pair–share

- Small group lab instruction with the teacher and students who need math support

---

Conclusion

In most instances, students cannot complete an inquiry activity without collecting, organizing, and interpreting data. Likewise, inquiry activities can require math skills such as measuring, counting, and computing in order for students to collect the data they need. Without question, math is an integral part of science. Science teachers should recognize the importance of math in the activities they plan for students. Science gives math relevance. It demonstrates to students that math matters. It gives students a purpose to use their math skills to some end, rather than as an end in themselves. As teachers plan inquiry activities, they should be cognizant of the math skills required, taking into consideration whether or not students have the math skills they need to complete the activity. If students do not have the required math skills, science teachers can support students carefully through their activity, spend time teaching the math needed as students complete the activity, or partner with the math teacher to provide connected instruction between students' math and science classes.

Stop and Reflect

1. How do the eight mathematical practices apply to what your students do when completing an inquiry activity in science?

2. What have your students done in science recently that required them to apply math skills? What was the activity, and which math skills did they use?

3. In considering the activity in question 2, were students proficient with the math skills they needed in order to successfully complete the activity? What did you do if they weren't? What would you have done if they weren't?

4. What digital tools can you use the next time students use math as a part of their science learning?

Chapter 8

Beyond the Classroom Walls

 "I never teach my pupils, I only provide the conditions in which they can learn."

—Albert Einstein

Science instruction should not be limited to four classroom walls, one teacher, and a specified number of students. In this world of virtual reality, science teachers can easily take their students on virtual field trips to see and experience people, places, and events related to any number of science concepts. Additionally, science teachers can bring in guest speakers (live, in-person, or digitally) to extend and clarify topics of study and interest. A study on middle school students' interest in science careers found that classrooms where teachers brought in guest speakers *and* used web-based science resources had a larger impact on student interest than those who did not use web resources (Koszalka, Grabowski, and Darling 2005). The suggestions in this chapter are intended to inspire teachers to reach beyond the confines of their classrooms and allow students to engage in a world of science exploration and learning.

Guest Speakers

Picture a scientist. Does he resemble Albert Einstein? Is he wearing a lab coat? When people draw pictures of scientists, they most often draw pictures of men with funny hair and in a lab setting with beakers and Bunsen burners bubbling over with some concoction. It's important for students to realize that scientists are human beings who may or may not exclusively conduct research in a laboratory setting. Girls and minorities, in particular, are influenced by meeting someone with whom they can personally relate when it comes to choosing science careers or being interested in studying science. Introducing young girls to women who have scientific careers provides a model for them to which they may themselves aspire. It makes them more aware of appropriate interests and pursuits for women, improves their attitudes toward the sciences, and may influence their choices in education and profession (Gray 2005).

Teachers can invite many different types of people into their classrooms: veterinarians, animal control officers, dentists, X-ray technicians, phlebotomists, nurses, forest rangers, fish and wildlife officers, forensic technicians, prosthetic designers/makers, scientific illustrators, engineers, science writers, meteorologists, greens-keepers, IT specialists, athletic trainers, and audiologists. People use different facets of science to do their everyday jobs in multiple fields. Being exposed to a wide range of career opportunities helps students see a stronger connection between what they learn in school and what they might need to do with that knowledge to have a career in science.

Guest speakers do not need to be professional speakers to address a group of students. Speaking to a group of students can be intimidating to some adults. However, guests should understand that they need simply tell students what they do on a typical day (in terms students can understand), discuss what they liked to do in school at that age, and talk a bit about what kinds of science they needed to learn to do their job. Demonstrations are not necessary, but any hands-on materials they can bring to show students something related to their work will make the presentation more interesting. Even having photos or specialized clothing makes an impression on students. Teachers should also try to arrange the speaker's topic to match with the curriculum being taught at that time. It is important to let the guest speakers know approximately how much time they have to talk, and leave time at the end for questions from the students.

Obtain an online video chatting account. Use an available computer camera to hold live video chats with scientists outside your immediate area. How far can your four classroom walls reach?

116

Benefits of Guest Speakers

Bringing guest speakers into the classroom throughout the year provides multiple benefits. According to Patricia Mullins, "A guest speaker should enhance the material you are covering" (Mullins 2001, 1). First of all, students are provided with invaluable career-planning advice. Most people who choose to talk to students about their professions do it because they love what they do, and are willing to share their avocations with others. This enthusiasm for their work also demonstrates to students that education can lead to a rewarding, fulfilling career. Exposing students to people who enjoy their careers is one way to motivate students and start them thinking beyond their school years. It also helps students see the connection between what they learn in science and how it can be used outside the classroom as a career choice.

Secondly, guest speakers enable schools to make connections with their communities. People who are not parents of current students, may not have a good picture of what goes on in an every day elementary- or middle-school classroom. Allowing community members to see a typical class and get a feeling for what teachers do everyday is good for public relations. Most non-educators who spend a few hours in a typical elementary- or middle-school classroom come away with a new appreciation for teachers and the work they do with students.

The Common Core State Standards Listening and Speaking Standard 3 includes specific language related to students' ability to ask and answer questions of speakers, and to summarize the key points, including details and examples from the speech (CCSSO 2010, 23). Science teachers can use guest speaker opportunities to help students learn and apply these critical skills. In preparation for the guest, students can collaborate to generate a list of essential questions they hope to have answered by the end of the presentation. Students can take notes, then identify and pose questions *relevant to the topic* that have not yet been answered. Common Core Writing Standard 4 requires students to produce writing appropriate to task and purpose (CCSSO 2010, 43). A nice gesture after the guest speaker's visit is to have students write thank-you notes in which students specify what they liked or learned from the presentation, citing specific information from the presentation.

Finally, a practicing specialist may be used as a source of information about a science area for which the class needs accurate, up-to-date information. A fish and wildlife officer, for instance, can provide teachers with resources about endangered species, pollution concerns, conservation efforts, and other environmental issues. Teachers may even be able to start professional relationships that will last for years, building portfolios of expert advisors. It's even possible that a visiting scientist might be able to find discarded or unused science equipment to donate to the classroom or the science department. These individuals might be able to support teachers wanting to find out about summer science programs in the area, professional development programs for teachers, or sponsored science camps for students. If these science specialists understand that teachers are committed to providing a rigorous science education for their students, they are more likely to offer help or advice when contacted.

Teachers who teach science to multiple classes have several options for bringing in guest speakers. They can have the speaker address just one or two classes while the teacher video records the session. For subsequent classes, students can email the speaker with their specific follow-up questions. Teachers might choose to try to set up the speaker as a special event and have all students in the grade or department join the session. To alleviate the interruption to instruction, the teacher setting up the speaker can collaborate with teachers of other subjects to generate instructional plans to support their content area. For example, the language arts teacher might assign a follow-up writing activity, the math teacher may use data charts related to the speaker's topic, the reading teacher may assign informational follow-up reading, and the social studies teacher may delve into the social, economic, or legislative effects of a topic related to the speaker's career.

Get Out There: Field Trips Worth Exploring

Field trips generate excitement in any classroom, including the science classroom. It is time away from the rigors and confines of school with peers. Many places offer learning experiences that students will remember for years to come. With a growing emphasis on STEM careers, even a nearby theme park may have special programs of which students and teachers may take advantage. Most amusement parks have educational programs, and special events and opportunities for teachers, students, and the community. However, lesser-known, more local area attractions might offer unique field trip experiences as well. T.I.G.E.R.S. Preservation Station at Barefoot Landing in North Myrtle Beach, South Carolina, has a wildlife exhibit and tiger museum. Here, students can have interactive experiences with rare and unusual animals. Project Wildlife in San Diego, California, offers demonstrations in schools or classrooms with specific animals to educate the public on protecting the wildlife. Project Wildlife's goal "is to improve the quality of life for local wildlife and the community as the primary resource for animal rehabilitation and conservation education" (2014, under "About Us").

In addition to theme parks and local attractions, many rural areas offer nature walks, bird-watching trails, and flower gardens. Students can go on scavenger hunts looking for evidence of specific wildlife, take pictures using hand-held devices (or digital cameras) to use in a slide show presentation about nature or living/nonliving things, or collect natural debris such as feathers, leaves, soil, or other small objects to look at under a microscope back in the classroom.

Students can visit any number of places where science is occurring all day, every day. These include, but are not limited to the following:

- power plant
- plant nursery
- fish hatchery
- working farm
- hands-on science museum
- orchard (or other area to personally pick produce)
- zoo
- hospital
- bird sanctuary
- natural history museum
- factory

Oftentimes, these types of locations provide ready-made lessons to use before, during, and/or after the trip. When used, students understand that the trip is not the end in itself; it is a means to learn interesting information in an alternative setting.

Of course, field trips do not necessarily need to require permission slips, buses, and chaperones. Some of the best science learning is already waiting for students in their very own schoolyards. In a learning journal, students can illustrate and describe plant and animal life that surrounds their school. They can catch bugs and observe them using hand lenses, then illustrate and describe them in their journal. They can collect rocks and minerals to sort and identify as one of the three types of rocks, or conduct simple tests to determine the minerals' hardness, color, luster, cleavage, and streak color. Science teachers could even bring students to a place in a nature area to read a story or informational text about something in nature.

Virtual Field Trips

Any time science teachers can bring students to a marine lab, a hands-on science lab, a water treatment plant, a natural landform, or any other science-related location, they demonstrate the reality of science in our everyday lives. They also expose students to career options that students might not ever have considered. However, the reality of school budgetary cuts and our national, state, and local economies might make real trips out of reach. Fortunately, today's science teachers have numerous other options to bring the outside world into their classrooms. Virtual field trips provide students with all the benefits of exploring the outside world within the confines of their four classroom walls. And students may explore on their own time, not having to rush or follow a tour guide's lead or race back to school to make the end-of-the-day bell.

Virtual Field Trip Benefits

Virtual trips benefit students in all settings and instructional situations, regardless of background, by increasing the experiential base of learners. An increase of the number of experiences directly adds to students' knowledge of content that they encounter in school (Marzano 2004). Virtual field trips extend students' experiences and provide a unique look at the world from a different perspective. Additionally, teachers can help build background knowledge for students who may not have advantages with science-related topics. Students who live on a farm may have real-life experiences with animals, whereas students who live in urban settings may never have seen a frog in its natural habitat. Students who frequent zoos, aquariums, and museums have an advantage over students who do not when learning about animal characteristics or animal habitats. Our everyday experiences shape our future learning. They provide us with an anchor upon which to build our existing knowledge and understanding. The more science teachers can expose their students to the world outside the classroom, the more knowledge they can gain.

Some virtual field trips are "slide show" in nature. They take students through a tour of a topic with text, images, diagrams, activities, and quizzes. Some slides provide multi-media and/or interactive features; some do not. Some virtual trips are video summaries, including narrations, images, demonstrations, and full explanations. To find a virtual field trip on a science topic of choice, simply use a favorite search engine. Type the topic name and *virtual field trip*. Find one that meets the information, presentation, and time requirements to support a topic or lesson. For example, teachers can take students on a virtual trip of New York City's Natural Museum of Natural History, watch a live webcam of Yellowstone National Park, or see what is happening live in space at NASA. The Jason Project is a "non-profit organization that connects students with scientists and researchers in real- or near-real time, virtually and physically, to provide mentored, authentic and enriching science learning experiences" (Jason Project 2014, under "homepage"). Here, students may (virtually) go on live expeditions with researchers and other scientists. Students who miss the live interactions may still view past expeditions or researcher interviews; they just will not have the advantage of the live interaction.

Once students experience the value of a virtual field trip, they may turn an open-ended project into one of their own. Having students use technology to create a virtual trip about a specific ecosystem, a true-to-life situation in their community, a how-to video explaining how light interacts with different mediums, or a microscopic trip through a common setting (such as the school cafeteria) will motivate students to not only learn the content associated with their project, but also allow for freedom of expression and creativity in informing others.

Outreach Programs

When science teachers cannot take students to off-campus locations, they can bring the off-campus locations to their students. Mobile science labs are privately owned and operated science outreach programs housed in trailers or other vehicles. Companies bring these outreach opportunities to schools. Inside the compartments are hands-on, exploration, and/or informational presentations in which students may participate. Some mobile labs operate with limited staff, therefore, the science teacher uses the equipment and technology in the lab to conduct any one or more of the prepared lessons and activities. Other labs come fully equipped with trained staff. They may make presentations, demonstrate science concepts, or monitor students as they themselves participate in hands-on activities. Examples include technology labs, biology labs, marine labs, agriculture labs, and food-science labs. Teachers can search for mobile science labs in their area through a general search, or they may try searching nearby colleges, universities, learning centers, or science organizations (e.g., aquariums and zoos) to see if something like this is available. These labs tend to come with a cost. However, they generally are much more cost-effective than bringing classes of students off campus for a limited time. This unique means of providing students with relevant, meaningful, and memorable science learning is worth the search to see what, if any, mobile science labs are available.

Students may also benefit from outreach programs provided by city, county, and state parks. Some parks are actually working parks. For example, the Crystal River Archaeological State Park in connection with the Florida Public Archaeology Network has, among other outreach programs, lab volunteer days. Here, people who may be interested in volunteer fieldwork can participate in actual archaeology digs along with trained professionals and university students. The Delaware State Park system has several outreach and field trip opportunities from which to choose with attention to specific grade level science topics.

Teachers should have a plan to hold students accountable for the information they learn during a guest speaker presentation, or an off-campus or virtual field trip. Special events are not a "free pass" for the day. Students can plan questions to ask or complete an interest inventory before the event, record learning on a graphic organizer or science journal during the event, and/or follow up with a summary of the information after the event.

Teachers might also consider looking into their local university system. For example, the University of Minnesota College of Science and Engineering sponsors outreach programs in STEM. Purdue University has a Discovery Learning Research Center that includes information about the annual Bug Bowl. The University of California, Los Angeles (UCLA) has several K–12 outreach programs related to science topics, including physics and astronomy, planetariums, as well as the OceanGLOBE program. This program is a beach research and outdoor environmental education program for upper elementary, middle- and high-school students.

Science Competitions

Teachers can do a great deal to support learning for students who struggle with science concepts (e.g., providing graphic organizers, outlines, etc.) to help students conceptualize the ideas presented, and offer a means for them to take notes and record their thinking and learning. (For information on differentiation strategies, see Chapter 5.) For teachers looking to enhance, extend, and challenge independent, critical thinkers, they could explore the idea of having students participate in special science-related competitions. Many competitions have categories for varied grade levels and each type of competition has its own guidelines. In general, students who enter science competitions investigate present-day issues or challenges, conduct and compile research results, and develop products or plan solutions to resolve problems. Some competitions have prizes just for entering; others have prizes only for the top winners. Teachers who guide students toward science topics that are appropriate for classroom curriculum can meet the standards while encouraging budding scientists to conduct authentic research and tackle real-world problems. Many competitions require teacher or other adult sponsors. Teachers may need to invest some time during the school day mentoring students and checking their progress.

One example of a science competition is called ExploraVision. This competition has been in place since 1992. Teacher sponsors guide students to identify a technology of interest, and then help students work collaboratively to develop a project detailing how this particular technology might look in 20 years. Individual students or student groups register, then develop and submit their project for the competition. Projects must follow a specific set of criteria. Once reviewed, teams may win savings bonds, a trip to Washington D.C., or any number of prizes provided by the sponsors. This competition has a deadline for entry and strict timelines for completion.

ExploraVision science competition encourages K–12 students to imagine what technology might be like in the future. ExploraVision helps teacher sponsors meet many of the National Science Education Standards while letting students experience scientific process and discovery in an engaging, hands-on way (ExploraVision 2014).

The Kids Science Challenge is another competition intended for students in grades 3, 4, 5, and 6. It is a free, nationwide science competition funded by the National Science Foundation (NSF) along with other foundations and corporations. Students design and submit innovative ideas related to each year's topics. Students can win a trip to work with a real scientist. Other participants may receive a science kit.

The Young Scientist Challenge is a science competition for middle school students in grades 5 through 8. Students can win cash, trips, and other prizes. Following competition rules and deadlines, students are judged on their scientific problem solving and communication skills. "The Young Scientist Challenge is designed to encourage the exploration of science and innovation among America's youth and to promote the importance of science communication. Over the past eleven years, more than 600,000 middle school students have been nominated to participate in the competition, and winners have gone on to speak in front of members of Congress, work with the nation's top scientists and pursue academic careers in the sciences" (Young Scientist Challenge 2014, under "About").

The Dupont Challenge, sponsored by numerous organizations and corporations, requires research, critical thinking, and science essay writing as students share innovative ideas related to urgent global challenges, and applied science, technology, engineering, and mathematics in daily living. Students choose a challenge and then follow the website's tips and guidance to complete a winning project. Both students and teachers win awards and are nationally recognized for their outstanding efforts. There are many additional science competitions to investigate including the following:

- National Engineers Week Future City Competition
- Odyssey of the Mind
- Intel® Science Talent Search
- Siemens Westinghouse Competition

Conclusion

Science happens every day, all day, all around us. Students can benefit from seeing, hearing, touching, and feeling experiences that take place outside their classroom walls. Teachers can set up informational talks with guest speakers. During these events, students learn about how science applies to the world around them. Science teachers might also want to explore the idea of taking a real or virtual field trip with their students. These adventures demonstrate first-hand how people use science in places out and about in their community. Finally, students can explore the world of science by participating in science-related competitions. Students who are highly competitive can apply their creative thinking, analytical processing, and problem-solving skills to attend to real-world tasks with real-world solutions.

Stop and Reflect

1. Why are learning extensions that reach beyond the classroom walls beneficial to students' overall science experiences and learning?

2. What do you see as the greatest challenges you face when extending learning beyond the classroom? Brainstorm some solutions to help ease these burdens.

3. Plan a day for a guest speaker to visit the class. Try to connect the topic to what students will be learning at that point in the school year. Plan a written follow-up activity for students after the presenter's visit.

4. Find an online field trip to use with students during an upcoming unit of study. Identify the topic and decide how you will hold students accountable for the information presented. Record the website and initial ideas on how to utilize it with students (and at what point in the learning process you will use it).

Chapter 9

Planning a Science Fair

 "All life is an experiment. The more experiments you make the better."

—Ralph Waldo Emerson

One of the most rewarding aspects of teaching science is seeing the pride on students' faces when they bring in their science fair projects. Christopher Gould, the former chairman of the California State Science Fair states "I believe the best way to encourage student accomplishment is to recognize it, and recognize it publicly" (Education World® 2014, under "How to Put on a Great Science Fair"). When organized successfully, students will have completed meaningful, relevant, interesting, and thoughtful projects, complete with scientific questions, data-driven investigations, and conclusions based on evidence and facts that support the learning goals of the science curriculum. Science fairs can require a great deal of time and effort on the part of both the teachers who organize them and the students who complete them. Teachers and students will need to commit to the time and effort required to have a successful fair. As students move through their educational careers, they will likely remember their science-fair projects and the results. If the project is structured in such a way that students use every aspect of scientific inquiry, they will remember the process long after the projects have been completed. Hopefully, the work they do to complete a project will inspire them to pursue additional topics and inquiries, leading to life-long inquiries.

Facts about Science Fairs

In some movie and television portrayals, science fairs typically amount to home-grown, exploding projects, illustrations, or demonstrations. If teachers were to poll their students as to the type of project they would likely complete right now, students may easily answer that they would create a model of the solar system or bring in a paper mâché or plaster of Paris erupting volcano. In fact, these are not the only types of science fair projects that students complete. Instead, students engage in open inquiry when they complete a science-fair project. This is where students generate a testable question, design and plan their own investigation, collect data and evidence, and report their results and findings. Usually, students display their science-fair project on a tri-fold project board. This board usually has a specific layout, and includes a summary of all of the components of a true scientific inquiry as shown in Figure 9.1.

Figure 9.1 Sample Science-Fair Display Board

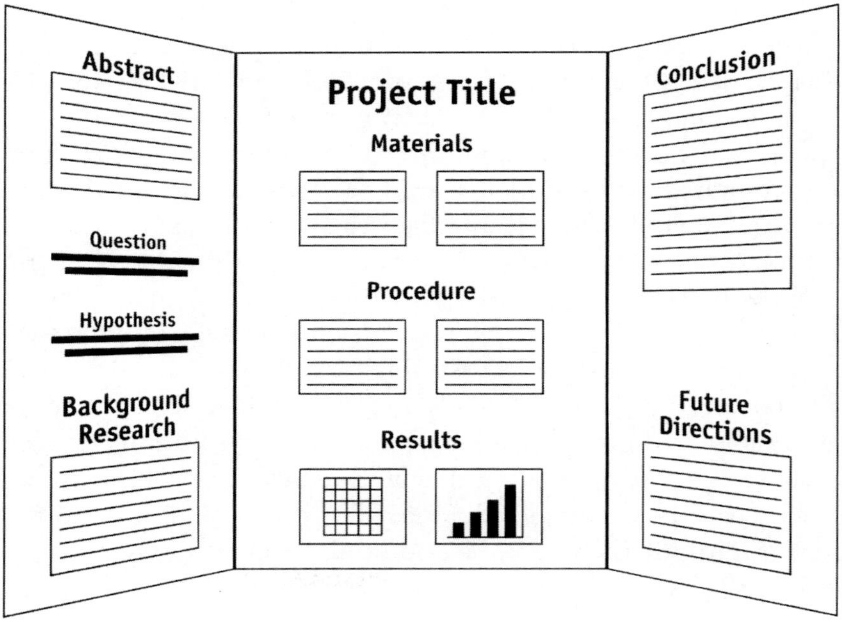

Science fairs offer students a creative freedom with their investigative talents. Although their teachers and their parents have supported and guided them through the process from beginning to end, students can take credit for the work, the research, and the product of their efforts. This can give them a great sense of accomplishment and confidence. However, it all begins in the classroom. Robert Marzano and Debra Pickering remind us that in order for students to engage in a task, they will first determine how they feel about it. Some of these feelings may be reflective of their teacher's feelings. Teachers who are energetic, enthusiastic, and committed to the process will have students who are energetic, enthusiastic, and committed to the process. "A teacher's intensity and enthusiasm are contagious and can have a positive effect on students' levels of attention" (2010, 30).

For more information on the levels of inquiry, see Chapter 4.

Considerations for Organizing Science Fair Projects

- Science fair projects should allow students to participate in open inquiries.

- Science fairs offer an opportunity for students, parents, teachers, administrators, and the area community to collaborate on and celebrate a meaningful event.

- Since some (or most) of the work may be done at home, teachers need a system to verify that the work is being done by students.

- Science fairs are time-consuming (but very rewarding) for fair coordinators, science teachers, and students.

- Two heads are better than one when organizing a school fair. Fair coordinators who use the buddy system will likely minimize tasks and pressure on themselves.

Beginning the Fair

The first step in the planning process is to know the *who, when, where, what, how,* and *why* of the fair.

1. Decide *who* will participate. Decide whether the fair will be required for all students at a particular grade level, or if students may participate on a volunteer basis. Science fairs typically need judges, organizers, and greeters. All volunteers should know their tasks and timeline for completion. Friendly reminders to all involved parties leading up to the fair keep everyone on track.

2. Decide *when* the fair will take place. It could take place during the school day, in the evening, or even on a weekend. The number of participants a school has will determine the duration of the fair. Most fairs last one or two hours. It is also important to establish timelines for participation. Students might need to complete a proposal to enter their project. These would need a specific deadline. Likewise, students should know the date their projects are due, when they will be judged, and when they may present them during the fair.

3. Decide *where* the fair will take place. Usually science fairs take place at the school. However, wherever the location, this space should be set up as far in advance of the fair as possible.

4. Think about *what* equipment or materials the fair requires. Tables, chairs, clipboards and pencils (for judges), water, and decorations are some materials to use during the fair. Also, the school may wish to support students' participation by providing science-fair boards to those who cannot acquire one. Most fairs distribute certificates and/or ribbons to participants, and sometimes trophies. Some schools, depending on their size, may wish to provide refreshments to visitors and participants. This requires additional planning and organization. Remember, volunteers could be put in charge of obtaining, setting up, distributing refreshments, and cleaning up the area after the fair is over.

5. Every detail about a science fair is important. Fair coordinators need to consider many details as to *how* the fair will run. How will fair information, including the timeline, be shared with students and parents? How long will the fair last? How will students bring and retrieve their projects before and after the fair? How will tables be set up? How will projects be organized and displayed? How will traffic flow as people visit each project? How will judging take place? How will students receive their awards? Each of these questions requires careful consideration. For example, fairs can be quite competitive. Coordinators may wish to establish a number system for the projects, and require that students' names and faces not show anywhere on the display. Then, when judges arrive, they may evaluate the project truly on its merit, eliminating the chance that a judge kindly gives extra points to someone he or she knows. Judging may include a criterion for how well the student explains his or her project. In this case, students would need to be present during judging, and the anonymity coding system is not needed. Some fairs organize the projects based on topic (e.g., all consumer projects together, all life science projects together). This provides for organization on fair day, and students and parents can see all the projects in one area that relate to theirs.

6. Science fair coordinators should ask themselves, "*Why* are we doing this?" Fairs are exhausting work, especially on a large scale. The real reason teachers do anything, of course, is for the students. It is all well worth it in the end.

Beginning the Project

Just as science fair coordinators need adequate time to organize, plan, and prepare for the fair, students need adequate time to plan, execute, and summarize their fair projects. Teachers should give as much lead-time as possible for students to begin the development of their project. First, teachers need to determine whether the fair will be required and whether students may work with a partner or small group. In real life, people in many professions collaborate with others to create products and to solve problems. Teachers in younger grades may opt to complete a class project. One of the first steps after these decisions are made is to provide students with a reasonable timeline for each stage of the project's completion. A sample timeline is provided in Appendix D. Then, the classroom teacher should provide consultation time with students or student groups during class to ensure that they have a testable question (teacher-approved), followed by a realistic procedure. Each investigation should lend itself to collecting measurable data, which may be plotted on a chart or graph. Students should receive a copy of the evaluation criteria before beginning work. A sample science fair project evaluation scale is available in Appendix D.

Some students or student groups may feel more comfortable investigating something they have studied during the school year or in past years. Others may want to explore ideas unrelated to school topics or ones that they find interesting. Either way, the initial inquiry should be student-generated. Some students may need more direct guidance with this step. Younger students may need more structure for science-fair projects. Teachers could supply a list of acceptable inquiries from which students or student groups may choose.

Students may need special equipment, such as scales and test tubes, which they probably do not have at home. Once all the project proposals are approved, the teacher can begin to assemble special material resources that students might need to be successful with their project, and plan class time for students to set up their experiments. At this point in the project, teachers could allow students to conduct their investigations at home. Some projects require students to collect data over several days or weeks, and in settings such as refrigeration or direct sunlight that may not be manageable during school hours.

Monitoring the Project

Once students are working on their projects, teachers should provide some time in class to check in, conference, and discuss their progress according to the timeline. Having students complete most of the work in the classroom requires adjusting the size of the space available for investigations, providing the necessary materials, and adjusting instruction to give ample time for students to complete the work. Teachers can plan supplemental activities (e.g., activity sheets, Internet-based activities, reading assignments, science logs) for students to complete when they are waiting for results or waiting for materials for initial setup. Students who keep up with their timelines will better stay on track to complete their projects on time.

Once the investigation has been completed, teachers may need to support students as they chart or graph the data, summarize components, and organize them on their fair boards. This will also allow teachers to guide students through the revision process, allowing students time to fix spelling errors, elaborate or clarify where needed, or redo data charts or graphs. If projects were completed at home, teachers can ask direct questions about the inquiry and procedures, ensuring that students understand the purpose of their investigation and the results.

Evaluating the Project

Once the projects have been completed, teachers may choose to evaluate the projects for a grade. Grading may take place before the actual fair. During the fair, students' projects are judged based on specific criteria. Teachers can use the same or similar criteria for grading, but this should not change the judge's marks. Grading serves an instructional purpose while evaluations by judges serve to identify the top projects among the student body. A science fair project rubric is one means of grading and/or evaluating projects. It spells out for students (and parents) what the criteria are for an outstanding project. Teachers can use the rating scale provided in Appendix D, create one using a word processing program better suited to the criteria they wish to emphasize, or use simple online rubric creators, such as RubiStar© or eRubric Assistant to set up a rubric using criteria important to them.

For students who participated in a paired or group project, teachers can modify the evaluation to include both independent and group criteria. Teachers might also require all the students in the group to write their own individual summaries of the project, including detailed descriptions of how they went about completing each step. This also acts as a sound writing activity, connecting sequenced essay writing into science class.

Day of the Science Fair

Likely, students are excited, anxious, and jittery on the day of the science fair. They have put a great deal of time and effort to construct and complete a project and compile the pieces into a striking display. Science-fair coordinators should have previously established a system for students to bring their projects for setup. If the fair takes place outside of the school, students should know when and where to be to set up their project. Once the projects are safely stored in their proper fair location, just the final details remain.

Students should understand the behavior expectations during a science fair. Fair coordinators may wish to meet with students immediately before the fair to review the school-wide expectations and consequences for non-compliance. Students should also understand their role during the fair. Should they stand beside their projects or can they walk around to view the other projects?

Judging may take place prior to the fair or during the fair itself. This step of the fair process usually takes longer than anticipated. Assigning more than 25 projects to each judge may lead to lagging results. This is one advantage to having judging take place before the actual fair. In this case, fair coordinators have adequate time to tabulate the results and determine the award recipients. For example, judging may take place on a Tuesday during the school day, but the school fair doesn't take place until Thursday evening. If judging does take place during the actual fair, coordinators should have a plan for award distribution, should the results be ready that night, or if they need more time.

The following are extra items fair coordinators might wish to have on hand during the actual fair:

- microphone or sound system
- tape (masking, clear, packing, or duct)
- permanent markers, pens, pencils, and clipboards
- scissors
- name tags (for teachers and volunteers, and perhaps students)
- brooms, dustpans, and trash receptacles
- cell phone or school radio

After the Fair

Fair coordinators should have previously established a plan for students to retrieve their projects. If some students go onto a district or state fair, their projects will need to be stored. Fair coordinators should also have a plan for any projects that are not picked up.

To facilitate an even more successful fair the following year, fair coordinators may have students and teachers write down any suggestions for improvement.

The final piece is the awards ceremony, assuming awards were not distributed during the actual fair. This can be accomplished during the school day. Students might come to a central location for their awards or have a special assembly. Fair coordinators may also wish to hold a separate, special awards event, similar to honor roll or a sports banquet for just this purpose.

Stop and Reflect

1. Brainstorm a few areas within your science curriculum where you could implement a science-fair project.

2. How will you grade students' science-fair projects? Create a grading rubric for this purpose.

3. How will you create time in your day to have students work on their science-fair projects?

4. How can you encourage parents/families to volunteer at a science fair?

Chapter 10

Assessment in the Inquiry-Based Science Classroom

 "The important question is not how assessment is defined but whether assessment information is used."

—Palomba and Banta

No lesson plan is complete without a means of assessing whether students met the learning objectives for a particular standard. Teachers should not assume that just because they taught it, students have learned it. They also want to be sure that students' science understanding goes beyond factual knowledge and information. With an emphasis on performance standards and with the blending of three dimensions (science and engineering, crosscutting concepts, and disciplinary core ideas), science teachers must reconsider how they assess student learning to meet the Next Generation Science Standards. "It is essential to understand that the emphasis placed on a particular Science and Engineering Practice or Crosscutting Concept in a performance expectation is not intended to limit instruction but to make clear the intent of the assessments" (NGSS 2013b, 2). Assessment of student learning may be accomplished through many means and techniques. This chapter explores evaluation strategies and provides examples of rubrics and grading systems.

For information on the three dimensions of the Next Generation Science Standards, see Chapter 1.

Back to the Beginning

As with all effective curricula planning, inquiry-based science instruction needs well-developed student learning objectives. Once the objectives are complete teachers may identify how they wish to assess and to what extent learning has occurred. Additionally, learning objectives can be written on varying levels of depth. One method is by using Bloom's Taxonomy. An online summary by Richard C. Overbaugh and Lynn Schultz explains the history of Bloom's. In 1956, Benjamin Bloom led a group of educational psychologists. They developed a classification system, in levels, of intellectual behaviors based on student learning. During the 1990s, Lorin Anderson (a former student of Bloom) led a new group of cognitive psychologists. They updated the taxonomy to reflect relevance to 21st century skills. Each category of Bloom's Taxonomy can be thought of as a step toward a deeper, more thorough understanding of a concept or scientific theory.

A sample learning objective written for students in grades 5 or 6 reads: Students will identify three desert plants or animals and give three details of each to explain how it is adapted to survive in the desert. This is a very specific, measurable learning objective but it is written at the knowledge/comprehension (lowest) level of Bloom's Taxonomy. Although the teacher may know whether students have a basic understanding of plant and animal adaptations, he or she will not be able to evaluate the depth of understanding students have (or don't have) with regard to the learning objective. The following is an example of how this learning could be revised to be at a higher level of Bloom's Taxonomy.

Apply/Analyze: Students will identify one non-desert-dwelling plant or animal and give three reasons to explain why it could not survive in a desert climate.

Evaluate/Create: Students will illustrate a fictional desert plant or animal adapted to live in this environment. Students will identify and explain three distinguishing characteristics that allow it to survive in a desert climate.

The apply/analyze objective in the previous example appears to include a similar verb stem to the knowledge/comprehension verb stem (identify). However, to meet this objective, students must distinguish characteristics and explain or rationalize the non-example. The evaluate/create learning objective requires more than students simply drawing a picture. They must also justify its characteristics. These two learning objectives, along with the original objective, allow teachers to more carefully determine who truly understands the relationship between animals and their habitats.

Summative Assessment Options

Summative assessments are those that evaluate student learning and are based on learning objectives. These are products or evidence of student learning that teachers typically use to grade students. Summative assessments generally come at the end of a unit of study, but this is not always the case. Teachers who give periodic quizzes use summative assessments during the learning cycle. Summative assessments are comprehensive; they help teachers see the whole learning picture of the student. Although tests may come to mind when thinking about summative assessments, teachers can use other methods to determine what, and to what extent, students mastered the learning objective. "Although higher-level skills, like critical thinking and analysis, can be assessed with well-designed multiple-choice tests, a truly rich assessment system would go beyond multiple-choice testing and include measures that encourage creativity, show how students arrived at answers, and even allow for collaboration" (Rotherham and Willingham 2009, 20). Science, perhaps more than any other subject, is the perfect content area for utilizing more advanced assessment options. Two such avenues include performance-based assessments and project-based assessments.

Performance-Based Assessments

One summative assessment method is to have students conduct a performance task. Performance-based assessments go beyond discovering if students *know* the science, but whether they can *do* the science. These assessments include having students manipulate equipment for a specific purpose, develop and/or conduct a mini-inquiry or part of an inquiry, or manipulate variables related to the topic. For example, to stay with the fifth- and sixth-grade animal adaptations goal, students might be provided a model or illustration of an animal and manipulate its characteristics to allow it to adapt to a desert environment. Students might re-size the animal (since desert dwellers tend to be small), give it sharp claws to allow it to burrow, color it a boring beige color to blend into its surroundings (keeping it safe from predators and allowing it to sneak up on prey), or give it long appendages as a cooling mechanism.

Performance-based assessments have greater reaching objectives than just assessing students' science learning. They have been connected to determining students' abilities to attend to 21st century skills as well. Robert Marzano and Tammy Heflebower (2011) identify 21st century skills as analyzing and utilizing information, addressing complex problems and issues, creating patterns and mental models, understanding and controlling oneself, and understanding and interacting with others.

Consider a performance task that requires students to research various desert animals from around the world. Students identify two of the same types of animals (e.g., two different types of scorpions, spiders, rabbits, eagles, coyotes) that live in two completely different desert areas. Students compare their characteristics and behavioral traits that allow them to survive in their desert habitat. Students might respond to the question, "Could the one animal live in the other's desert region? Justify your thoughts." When considering the requirements set before students to complete this performance task, first, students must be able to apply research skills effectively. They must be able to synthesize specific yet related information about two animals from two different deserts and apply advanced analytical skills. This task may seem complex and daunting; yet, it is what today's science students must be able to do to succeed as 21st century learners. By pairing students for this performance task, teachers can add the dimension of collaboration to the assessment as they evaluate student learning. This performance task then becomes a much more far-reaching, broader, yet scientifically purposeful assignment for students as they demonstrate their understanding of this goal.

Help for Finding Appropriate Performance Task Items

Teachers can access Performance Assessment Links in Science (PALS), a free online standards-based website devoted to science performance task items. This site is grant-funded through the National Science Foundation. It includes K–12 links to the National Science Education Standards, performance tasks by content, directions for teachers and students, evaluation rubrics, and student examples.

One important element related to assessing students with performance tasks is to have a quality rubric or evaluation system in place ahead of time. By carefully constructing evaluation criteria upon which to determine the students' depth of understanding, teachers can determine who has met this goal completely, who has not met this goal at all, and can identify which students fall somewhere in between.

Performance-Based Assessments

Another method of assessing students is closely related to performance tasks. Performance-based assessments allow students to create a project related to the learning goal. This type of assessment requires students to *do* something with the science they learned. The project should require "critical thinking, problem-solving, collaboration, and various forms of communication" (Solis and Larmer 2012, 50). One might argue that if they *do* something, they *make* something. Therefore, there is a fine distinction between the performance (doing) and the project (making). Whatever teachers call it, the idea is that students are offered an opportunity to combine their learning with creative thinking in a meaningful format to demonstrate understanding.

Students tend to like making posters, building models, or creating slide shows. However, teachers should be aware that projects that only require students to regurgitate information in a creative display or presentation are really only attending to the lowest levels of Bloom's Taxonomy (remembering and understanding). With the Internet so full of facts, students can download information and incorporate it into a display without even having to read it. The teacher only learns how well students can search, find, and print information off the Internet. Although Internet research is an important 21st century skill, it does not necessarily assess student learning of the science topic.

Elevating projects to the application and higher levels of Bloom's Taxonomy requires just a few simple alterations to the project requirements. For example, Mrs. K wants her students to create an interactive slide show presentation about a lesser-known desert animal. Included in the information must be three or more adaptations the animal has to live in its desert environment. This is a common, simple, straightforward project. Students would learn something new. However, they would not be applying information; they would just be restating information they read. Instead (or in addition), the teacher may have students write a story as if they were the lesser-known desert-dwelling animal, detailing how its adaptations allow it to survive in its habitat. Every story has a rising action, climax, and conclusion. In the desert, any number of events might take place to put the animal in danger. The animal must use a characteristic or behavioral adaptation to escape impending doom. With this project, students must consider the animal's survival from its own perspective. They must think creatively, applying what they know and learned in a unique situation. In this instance, students truly are creating something. They are, in essence, doing science.

Projects can take on any number of creative formats as shown in Figure 10.1. Teachers should consider the tasks they assign students, and strive to employ the higher levels of Bloom's Taxonomy to see a greater sense of student learning.

Figure 10.1 A Sample List of Projects to Assess Student Learning

• Book Jacket	• Instruction Manuals	• News Reports
• Brochures	• Interviews	• Poems
• Cartoons	• Inventions	• Recipes
• Debates	• Letters	• Scrapbooks
• Diagrams	• Lyrics	• STEM Activities
• Experiments	• Models	• Stories
• Games	• Multimedia Presentations	• Web Quests

Lab Reports and STEM

Teachers can use lab reports as a way to assess student learning with regard to a unit of study. Usually, labs are completed as a means of instruction, offering students hands-on experiences during the learning process. However, labs can also serve as summative performance tasks or project-based outcomes. This is especially true when students engage with Science, Technology, Engineering, and Mathematics (STEM) activities. These problem-based projects make for comprehensive, culminating activities. They require students to assimilate a multitude of information in a complex, organized activity at the creativity level of Bloom's Taxonomy.

> For information about STEM activities, see Chapter 4.

Lab reports are more than simple summaries of an inquiry activity. Teachers should require more of students than simply explaining what they learned when they conducted an investigation. Summarizing is an important writing skill, and science teachers can support literacy by requiring students to write summaries of their learning. To be more comprehensive and summative in nature, teachers might ask students to review all the investigations they conducted during this unit, select the one that truly demonstrated a specific goal, and explain why these results are more meaningful than other results. In this instance, students analyze all the investigations in which they took part, evaluate them on their merit, and justify their ideas regarding the "best" one.

Traditional Tests

Finally, as mentioned earlier, tests are common methods of assessing students' understanding. Tests have many advantages for both teachers and students. They are objective and take less time than performance tasks or projects. However, most test questions target only basic knowledge and comprehension levels. They usually do not require analytical thinking, nor do they require students to evaluate situations or ideas. However, teachers can make simple modifications to their tests to elevate the cognitive levels of the questions, and thus, student responses.

One way teachers can elevate the level of test questions is to have students justify their responses. For example, consider the following test question that could be presented on a fifth- or sixth-grade test about animal adaptations:

Sample Question: Why might a desert species, such as the Gila monster, have bright colors?

    A. They warn predators that they are poisonous.

    B. They warn prey that they are fast.

    C. They attract prey.

    D. They attract insects upon which the Gila monster feeds.

The correct answer is A. During their study, students would have learned that bright colors can warn predators that the species is dangerous. In the case of the Gila monster, its colors also camouflage it from predators. Assuming students did not study the Gila monster specifically, this question challenges students to apply their learning (bright color adaptations) in a new context. To elevate this question even more, the teacher might require students to justify why they selected their response, and/or justify why they did NOT select the other responses. Now, students must not only apply their understanding of this particular adaptation, they must also analyze and evaluate incorrect answer choices.

The previous example illustrates a slightly higher-level question than usual. As teachers review published assessments or create their own, they can embed questions that challenge students' thinking, and require higher levels of understanding. For example, consider this question:

Sample Question: Which animal's foot is best adapted for burrowing in desert soil?

    A. webbed toes

    B. sharp claws

    C. flat feet

    D. hooved feet

The correct answer is B. Burrowing animals need sharp claws to dig through sand (and other types of soil). Assuming teachers did not teach this particular adaptation directly, students must apply their understanding of desert animal adaptations to correctly attend to this question. To raise the cognitive level of the question, teachers can ask students to illustrate what a desert burrowing animal's foot might look like. This test question has been elevated from the *applying* level to the *creating* level. These simple modifications to standard test questions challenge students to think, apply, and justify their ideas, which are important skills beyond actually knowing facts related to the science content.

Another consideration is to include test items that require students to interpret tables, graphs, diagrams, or illustrations. These types of questions are higher on the Bloom's Taxonomy scale. Standardized tests tend to include a good sampling of these items. Teachers who include them in classroom assessments show students how to apply their quantitative literacy skills to solve real-world problems, as well as assess their abilities to analyze information.

Formative Assessment Options

According to Learning Point Associates®, "Formative assessment is a process in which teachers use various tools and strategies to determine what students know, identify gaps in understanding, and plan future instruction to improve learning" (Learning Point Associates 2009, 2). Therefore, the teacher should use formative assessment throughout the course of study to drive their instruction. If students are struggling with the content of a lesson, teachers can make timely changes to the instructional design to better support student learning. Formative assessments provide information about student learning for the teacher and also show students where they are along the learning path. Teachers might simply ask a question to which students respond (orally or in writing), or be more formal by providing a short quiz.

Observation

Probably the most common formative assessment technique is teacher observation. As teachers walk around the room, they make observations about student actions, written work, and comments. If a teacher notices that students are disengaged with a particular collaborative project, they can pose direct, probing questions to these students to redirect students to on-task behavior.

Perhaps during teachers' observations, they notice that not just one or two students are disengaged, but rather, a majority of the class seems confused or off-task, or they are completing the task incorrectly. At this point, the teacher would likely stop, redirect the attention of the whole class, and clarify the expectations. The teacher might even model one or two examples so that students may better follow the procedures. After the lesson, teachers may reflect on what did not go well, as well as what went well, using a checklist for themselves. This type of personal teaching journal will hopefully lead teachers to make adjustments or corrections the next time they set students to a particular task. Figures 10.2 and 10.3 provide examples of criteria that might be on a teacher's self-reflection checklist for two types of inquiry lessons: structured and open.

Figure 10.2 Sample Formative Assessment Checklist for Structured Inquiry Lessons

Item	Yes	No	Possible Changes
Are there enough materials?		✓	get more test tubes next time; need more paper towels
Do students understand the directions?		✓	rewrite; go over directions before beginning
Are students mostly on task?	✓		
Are students measuring correctly, using proper units?		✓	need to review metric rulers and centimeters before activity
Do students understand the main concept they are investigating?	✓		
Are students writing down their observations?		✓	some need several reminders
Are students doing the calculations correctly?		✓	need to review decimal places
Are student findings reasonable—do they match the lesson objectives?	✓		

Figure 10.3 Sample Formative Assessment Checklist for Open Inquiry Lessons

Item	Yes	No	Possible Changes
Can students devise testable questions?	✓		some needed prompting; conduct additional small group activity: identifying testable and non-testable questions
Can students devise appropriate procedures to test their questions?		✓	needed guidance; conduct mini-lessons of this step
Can students accurately record observations and collect evidence?		✓	needed guidance; conduct mini-lessons of this step, especially graphing
Are students adequately managing their time?	✓		with reminders from teacher; set digital timer and remind students where they should be at various points during class
Did students clean up their materials and area?	✓		with reminders from teacher
Can students answer their questions based on evidence collected?	✓		
Can students explain the main scientific ideas in their own words?	✓		

Summaries

Summarizing is a writing skill required in the Common Core State Standards for English Language Arts beginning in fourth grade. Up until that point, students must recall information to answer a question in writing. Reading Rockets states, "Summarizing teaches students how to discern the most important ideas in a text, how to ignore irrelevant information, and how to integrate the central ideas in a meaningful way. Teaching students to summarize improves their memory for what is read. Summarization strategies can be used in almost every content area" (2014, under "Summarizing"). As learning progresses from one idea to the next, science teachers can pose informational questions for students to summarize. This might take the form of a one-minute or one-sentence summary, a response to the lesson's essential question, or a personal reflection on the learning in a science journal. Any of these activities may take place before, during, or after the lesson. An example of a summary strategy is exit cards. This could be used as a formative assessment strategy where the teacher poses a question about learning that should have occurred during class, and students respond on a note card (or slip of paper) before they exit the class. With a quick flip through students' responses, teachers can gauge how completely students understand the day's concept(s). The day's question might simply be, "What did you learn today?" It might be more focused on a particular skill or concept, such as, "What are three examples of desert animal adaptations?"

Technology and Formative Assessments

To incorporate technology, teachers could have students use their cell phones to text them a short summary. Other options include instant messaging or chatting on a website specifically created for an educational forum. Teachers interested in using this technology should check the protocols in their school or district before setting up accounts and interacting with students digitally.

Teachers who review their students' summaries before the next class period can learn which concepts, ideas, processes, and/or vocabulary words need to be reinforced. Teachers can also differentiate their science instruction, strategically placing students who need more focused instruction in a small, group that will work with the teacher the next day while the other students extend their learning independently or with a partner completing a reading, writing, research, or alternative hands-on project.

Evaluating Student Work

Assessment strategies should include a plan for quantifying student learning. Most science teachers are required to grade students. Grades are generally based on a pre-determined scale. Some assessments are easier to quantify than others. For example, if a test has 10 questions, each question would be worth 10 points on a 100 point scale. However, if teachers want to include written tasks along with the tests, they will need to devise a system to evaluate how well students performed on the written components and include this result in the overall grade. Let's say a teacher has students justify one of the 10 answer choices on the multiple choice test. He or she may assign up to 4 points for this justification using very simple criteria: (0) no response; (1) inadequate response; (2) partial response; (3) adequate response; and (4) complete/thorough response. Then, the teacher need only determine how these points will be added into the final test grade. Instead of a 10-question test, perhaps students are evaluated on a 14-point test. These points could simply be added to the grade based on the 10-point test. Perhaps the test will be based on 96 points, making each question worth 9.5 points (instead of ten), with the written piece added to make a full 100 points. Whichever technique teachers decide to use, students should know what the expectations are ahead of time.

For performance- and project-based assessments, teachers can use rubrics, rating scales, or checklists to evaluate student work. These evaluation tools can be used for nearly every type of project or written assessment. They also provide a clear-cut way to present the grading criteria to students and to parents. Rubrics include specific criteria for each numeric credit. Rating scales are a little more subjective. They include similar evaluation criteria as rubrics, but teachers assign a wider range of points for each criterion. Rubric-building websites allow teachers to create their own rubrics. They also provide multiple examples of rubrics for use in nearly every grade, subject, or category. If teachers share the grading rubric or scale with the students (and parents) at the onset of the project, students have the opportunity to understand the expectations and requirements to earn the expected letter grade. Checklists are most helpful when teachers evaluate students' process skills, or when teachers want students to self-evaluate their progress. Checklists provide a list of yes/no statements—either the student can do it, or they cannot do it.

Figure 10.4 shows a sample primary checklist related to physical science concepts. Figure 10.5 shows an example of a rating scale for evaluating a project. Teachers who use a 4.0 scale to grade students can determine an average score by dividing the students' total score by eight (the number of criteria). Teachers who use a 100-point scale can divide by the students' total score by 32, then multiply by 100. Several options related to project-based assessments were presented in Chapter 5. One idea mentioned was to have students design a new species that is uniquely adapted to the desert environment, and write a placard for the zoo where the animal is kept to explain where it lives and its adaptations for living there. Figure 10.6 illustrates a rubric a teacher might use to evaluate students' overall performance with this task. Notice that both evaluation models include criteria to evaluate students' writing skills and science learning. Full-size versions of Figures 10.4, 10.5, and 10.6 can be found in Appendix D.

Figure 10.4 Sample Primary Checklist for Physical Science Concepts

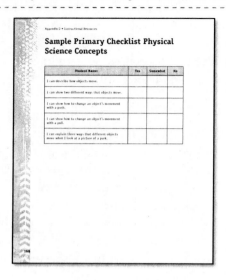

Figure 10.5 Sample Rating Scale for Animal Adaptation Story

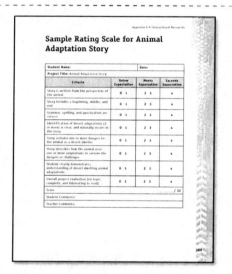

Figure 10.6 Sample Rubric for Unique Animal and Zoo Placard Summary

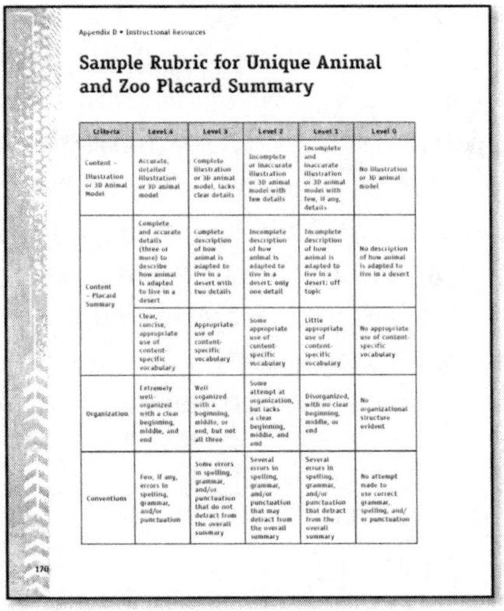

Appendix D • Instructional Resources

Sample Rubric for Unique Animal and Zoo Placard Summary

Grading Cooperative Learning Groups

Although many educators believe that a group task demands a group grade, this is not always the case. Speaking of Teaching states, "Grading the group achievement overall should be based both on the success of the final product and the group's assessment of its operations" (Stanford University 1999, 4). For example, there may be a student within a group who either refuses to cooperate and do his or her share. Other group members should not be penalized if one of the members does not participate within the group. Teachers can use a rubric that includes criteria related to how well each group member actively engaged in the activity. Conferencing with disruptive students, reviewing the expectations for group work, and/or devising individual behavior plans (with specific behavior goals) with these students can help some rethink their misbehavior before it happens.

Conclusion

The assessment component is an essential part of any science lesson planning process. Teachers have options when assessing student learning. Formative assessments determine where students are along the learning process. These results are usually not graded. Instead, they help teachers inform instruction. Depending on how student learning progresses, science teachers may re-teach concepts or extend learning opportunities at any point along the instructional path. Summative assessments allow science teachers to formally evaluate student learning. These may take the form of traditional tests, or they may offer students more creative means to demonstrate their learning through performance tasks and/or collaborative or independent projects. Teachers should utilize varied assessments to target not just students' knowledge and understanding of science topics, but also to allow students to demonstrate deeper levels of understanding through application, analysis, evaluation, and creativity tasks. Teachers who clearly inform students of the evaluation criteria regarding both formative and summative assessments best prepare students to succeed with their science learning.

Stop and Reflect

1. How are assessments related to learning goals or learning objectives?

2. How can teachers incorporate more than just tests using a summative assessment? Why is this important when determining how well a student learned the topic, skill, or process?

3. How do you use formative assessments to guide instruction? Provide an example of a time when a formative assessment measure redirected your lesson plans.

4. Create a rubric or rating scale to evaluate a summative project or performance task. Assign the project or task, and use the rubric or checklist to evaluate student learning. Then, make revisions, if needed, to improve the quality of the rubric or checklist.

References Cited

Achieve, Inc. 2013a. Next Generation Science Standards. "Executive Summary." http://www.nextgenscience.org/sites/ngss/files/Final%20Release%20NGSS%20Front%20Matter%20-%206.17.13%20Update_0.pdf.

Achieve, Inc. 2013b. Next Generation Science Standards. "Appendix A—Conceptual Shifts in the Next Generation Science Standards." http://www.nextgenscience.org/sites/ngss/files/Appendix%20A%20-%204.11.13%20Conceptual%20Shifts%20in%20the%20Next%20Generation%20Science%20Standards.pdf.

Achieve, Inc. 2013c. Next Generation Science Standards. "Appendix H—Understanding the Scientific Enterprise: The Nature of Science in the Next Generation Science Standards." http://www.nextgenscience.org/sites/ngss/files/Appendix%20H%20-%20The%20Nature%20of%20Science%20in%20the%20Next%20Generation%20Science%20Standards%204.15.13.pdf.

Achieve, Inc. 2013d. Next Generation Science Standards. "Appendix L—Connections to the Common Core State Standards for Mathematics." http://www.nextgenscience.org/sites/ngss/files/Appendix-L_CCSS%20Math%20Connections%2006_03_13.pdf.

Achieve, Inc. 2013e. Next Generation Science Standards. "Appendix M—Connections to the Common Core State Standards for Literacy in Science and Technical Subjects." http://www.nextgenscience.org/sites/ngss/files/Appendix%20M%20Connections%20to%20the%20CCSS%20for%20Literacy_06_03_13.pdf.

Achieve, Inc. 2013f. Next Generation Science Standards. "Three Dimensions." http://www.nextgenscience.org/three-dimensions.

Achieve, Inc. 2013g. Next Generation Science Standards. "The Next Generation Science Standards." http://www.nextgenscience.org/next-generation-science-standards.

Allen, Janet. 1999. *Words, Words, Words: Teaching Vocabulary in Grades 4–12.* Portland, ME: Stenhouse Publishers.

Anderson, Lorin W., and David R. Krathwohl, eds. 2001. *A Taxonomy for Learning, Teaching and Assessing: A Revision of Bloom's Taxonomy of Educational Objectives: Complete Edition.* New York, NY: Longman.

Banchi, Heather, and Randy Bell. 2008. "The Many Levels of Inquiry." *Science and Children* 6 (2): 26–29.

Barton, Mary Lee, and Deborah L. Jordan. 2001. *Teaching Reading in Science: A Supplement to Teaching Reading in the Content Areas Teacher's Manual,* 2nd ed. Aurora, CO: Mid-continent Research for Education and Learning.

Beers, Kylene. 2003. *When Kids Can't Read: What Teachers Can Do.* Portsmouth, NH: Heinemann.

Bloom, Benjamin S., and David R. Krathwohl, eds. 1956. *Taxonomy of Educational Objectives: The Classification of Educational Goals, Handbook 1: The Cognitive Domain.* New York, NY: David McKay.

Brummitt-Yale, Joelle. 2014. "The Relationship Between Reading and Writing." K12Reader. http://www.k12reader.com/the-relationship-between-reading-and-writing/.

Buehl, Doug. 2001. *Classroom Strategies for Interactive Learning,* 2nd ed. Newark, DE: International Reading Association.

Cech, Scott J. 2007. "Job Skills of the Future in Researchers' Crystal Ball." *Education Week.* http://www.edweek.org/ew/articles/2007/06/20/42skills.h26.html.

Chitman-Booker, Lakenna, and Kathleen Kopp. 2013. *The 5E's of Inquiry-Based Science.* Huntington Beach, CA: Shell Education.

Cooper, Donna, Adam Hersh, and Ann O'Leary. 2012. "The Competition that Really Matters." Center for American Progress. http://www.americanprogress.org/issues/economy/report/2012/08/21/11983/the-competition-that-really-matters/.

Curwood, Jen Scott. 2013. "Summer Trips for Teachers." *Scholastic.* http://www.scholastic.com/teachers/article/summer-trips-teachers.

Davis, Barbara Gross. 1999. "Cooperative Learning: Students Working in Small Groups." *Speaking of Teaching: Stanford University Newsletter on Teaching* 10 (2): 1-4.

Davis, Michelle R. 2007. "Digital Tools Push Math, Science to New Levels." *Education Week.* http://www.edweek.org/dd/articles/2007/06/20/01sr_curriculum.h01.html.

Duke, Nell, and P. David Pearson. 2002. "Effective Practices for Developing Reading Comprehension." In *What Research Has to Say About Reading Instruction*, edited by S. Jay Samuels, 205–242. Newark, DE: International Reading Association.

Dunne, Diane Weaver. 2000. *Education World*. "How to Put on a Great Science Fair!" http://www.educationworld.com/a_curr/curr220.shtml.

International ICT Literacy Panel. 2002. "Digital Transformation: A Framework for ICT Literacy." Educational Testing Services. http://www.ets.org/Media/Tests/Information_and_Communication_Technology_Literacy/ictreport.pdf.

ExploraVision. 2014. "About ExploraVision." Accessed July 23. http://www.exploravision.org/.

Frayer, Dorothy A., Wayne C. Frederick, and Herbert J. Klausmeier. 1969. "Working Paper No. 16: A Schema for Testing the Level of Concept Mastery." Madison, Wisconsin: Wisconsin Research and Development Center for Cognitive Learning.

Goodwin, Bryan, and Kristen Miller. 2013. "Technology-Rich Learning: Research Says/Evidence on Flipped Classrooms is Still Coming In." *Educational Leadership* 70 (6): 78–80.

Gray, Jennifer B. 2005. "Sugar and Spice and Science: Encouraging Girls through Media Mentoring." *Current Issues in Education* 8 (18).

Harvard-Smithsonian Center for Astrophysics, Science Education Department. 1987. "A Private Universe" (Video). Cambridge, MA: Science Media Group.

Harvey, Stephanie, and Anne Goudvis. 2000. *Strategies That Work: Teaching Comprehension to Enhance Understanding*. Portland, ME: Stenhouse Publishers.

Helios Education Foundation. 2013. "STEMming the Decline in Math and Science." http://www.helios.org/news-media-details.aspx?id=127.

Hossain, Mokter, and Michael G. Robinson. 2012. "How to Motivate US Students to Pursue STEM (Science, Technology, Engineering and Mathematics) Careers." *US-China Education Review* A 4: 442–451. http://files.eric.ed.gov/fulltext/ED533548.pdf.

Huffington Post. 2012. "U.S. Students Still Lag Behind Foreign Peers, Schools Make Little Progress in Improving Achievement." *Huffington Post*. http://www.huffingtonpost.com/2012/07/23/us-students-still-lag-beh_n_1695516.html.

Jason Project. 2014. "Homepage." Accessed July 23. http://www.jason.org/.

Johnson, Ben. 2011. "How to Creatively Integrate Science and Math." *Edutopia.* http://www.edutopia.org/blog/integrating-math-science-creatively-ben-johnson.

Kopp, Kathleen. 2010. *Everyday Content-Area Writing: Write-to-Learn Strategies for Grades 3–5.* Gainesville, FL: Maupin House Publishing, Inc.

Koppal, Mary. 2002. "Heavy Books Light on Learning: Not One Middle Grades Science Text Rated Satisfactory by AAAS's Project 2061." Project 2061. http://www.project2061.org/about/press/pr990928.htm.

Koskinen, Patricia S., and Irene H. Blum. 2006. "Paired Repeated Reading: A Classroom Strategy for Developing Fluent Reading." *The Reading Teacher* 40 (1): 70–75.

Koszalka, Tiffany A., Barbara L. Grabowski, and Nancy Darling. 2005. "Predictive Relationships between Web and Human Resource Use and Middle School Students' Interest in Science Careers: An Exploratory Analysis." *Journal of Career Development* 31 (3): 171–184.

LaValley, Brian. "SUNYCO's Heritage in Technology Education." Oswego State University of New York. Accessed July 2, 2014. http://www.oswego.edu/academics/colleges_and_departments/departments/technology/about_us/history.html.

Layton, Lindsey. 2013. "U.S. Students Lag Around Average on International Science, Math, and Reading Test." *Washington Post.* http://www.washingtonpost.com/local/education/us-students-lag-around-average-on-international-science-math-and-reading-test/2013/12/02/2e510f26-5b92-11e3-a49b-90a0e156254b_story.html.

Learning Point Associates. 2009. "Connecting Formative Assessment Research to Practice: An Introductory Guide for Educators." http://www.learningpt.org/pdfs/FormativeAssessment.pdf.

Marzano, Robert J. 2004. *Building Background Knowledge for Academic Achievement.* Alexandria, VA: Association for Supervision and Curriculum Development.

Marzano, Robert J., and Debra J. Pickering. 2010. *The Highly Engaged Classroom.* Bloomington, IN: Marzano Research Laboratory.

Marzano, Robert J., Debra Pickering, and Jane E. Pollock. 2001. *Classroom Instruction that Works: Research-Based Strategies for Increasing Student Achievement.* Alexandria, VA: Association for Supervision and Curriculum Development.

Marzano, Robert J., and Tammy Heflebower. 2011. *Teaching & Assessing 21st Century Skills.* Bloomington, IN: Marzano Research Laboratory.

McLeod, Joyce, Jan Fisher, and Ginny Hoover. 2003. "Chapter 2. Managing Instructional Time." ASCD. http://www.ascd.org/publications/books/103008/chapters/Managing-Instructional-Time.aspx.

Merriam-Webster Collegiate Dictionary. 2005. 11th ed., s.v. "inquiry."

Michigan Science Teachers Association. 2014. "Writing Across the Curriculum." Accessed July 2. http://www.michigan.gov/documents/mde/Science_WAC_2_3_264454_7.pdf.

Moore, David, Sharon Arthur Moore, Patricia M. Cunningham, and James W. Cunningham. 2006. *Developing Readers and Writers in the Content Areas K–12.* Boston, MA: Allyn & Bacon, Inc.

Mullins, Patricia A. 2001. "Using Outside Speakers in the Classroom." *Observer* 14 (8).

National Academy of Sciences. 2007. *Rising Above the Gathering Storm: Energizing and Employing America for a Brighter Economic Future.* Washington, DC: The National Academies Press.

National Governors Association (NGA) Center for Best Practices and Council of Chief State School Officers (CCSSO). 2010. "Common Core State Standards: English Language Arts Standards." Washington, DC: National Governors Association Center for Best Practices, Council of Chief State School Officers.

National Research Council. 1996. "National Science Education Standards." Washington, DC: National Academy Press.

National Research Council. 2012. *A Framework for K–12 Science Education: Practices, Crosscutting Concepts, and Core Ideas.* Washington D.C.: National Academies Press.

National Science Teachers Association. 2014. The NSTA Learning Center. Accessed July 2. http://learningcenter.nsta.org/.

National Writing Project. 2002. "Thinking About the Reading/Writing Connection with David Pearson." *The Voice* 7 (2).

North Central Regional Educational Laboratory. 2003. "enGauge 21st Century Skills for 21st Century Learners: Literacy in the Digital Age." http://pict.sdsu.edu/engauge21st.pdf.

Ogle, Donna, Ron Klemp, and Bill McBride. 2007. *Building Literacy in Social Studies.* Alexandria, VA: Association for Supervision & Curriculum Development.

Organization for Economic Development and Cooperation. 2014. Program for International Student Assessment (PISA). Accessed July 23. http://www.oecd.org/pisa/.

Overbaugh, Richard C., and Lynn Schultz. 2014. "Bloom's Taxonomy." Accessed August 5. http://ww2.odu.edu/educ/roverbau/blooms-taxonomy.htm.

Pauk, Walter. 2007. *How to Study in College*, 9th ed. Boston, MA: Houghton Mifflin.

Pinchok, Nick, and W. Christopher Brandt. 2009. "Connecting Formative Assessment Research to Practice: An Introductory Guide for Educators." Learning Point Associates. http://www.learningpt.org/pdfs/FormativeAssessment.pdf.

Pollack, Eileen. 2013. "Why Are There Still So Few Women in Science?" *New York Times*. http://www.nytimes.com/2013/10/06/magazine/why-are-there-still-so-few-women-in-science.html?pagewanted=all.

President's Council of Advisors on Science and Technology. 2010. "Prepare and Inspire: K–12 Education in Science, Technology, Engineering, and Math (STEM) for America's Future." Washington, D.C.

Project Wildlife. 2014. "About Us." Accessed June 27. http://www.projectwildlife.org/about.php.

Reading Rockets. 2014. "Summarizing." Accessed July 23. http://www.readingrockets.org/strategies/summarizing.

Rotherham, Andrew J., and David Willingham. 2009. "21st Century Skills: The Challenges Ahead." *Educational Leadership* 67 (1): 16–21.

Shemkovitz, Anne. 2014. "Helping Young Readers Navigate Nonfiction Texts." *ASCD Express*. Accessed June 28. http://www.ascd.org/ascd-express/vol4/414-newvoices.aspx.

Siegel, Efrem. 1963. "NDEA: Progress Report." *The Harvard Crimson*. http://www.thecrimson.com/article/1963/10/2/ndea-progress-report-peducators-who-are/.

Solis, Alfred, and John Larmer. 2012. *Project Based Learning Workbook Series: PBL 101 Workbook: The Companion to BIE's Introductory Project Based Learning Workshop,* 2nd ed. Novato, CA: Buck Institute for Education.

Stanford University. 1999. "Speaking of Teaching." *Stanford University Newsletter on Teaching* 10 (2).

Stanford University. 1999. "Cooperative Learning: Students Working in Small Groups." *Stanford University Newsletter on Teaching* 10 (2).

STEM Education Coalition. 2014. Accessed July 2. http://www.stemedcoalition.org/.

Stronge, James H., Pamela D. Tucker, and Jennifer L. Hindman. 2014. "Chapter 3. Classroom Management and Organization." Accessed June 20. ASCD. http://www.ascd.org/publications/books/104135/chapters/Classroom-Management-and-Organization/aspx.

Sunnyside Unified School District. 2014. "Sunnyside District." Accessed July 23. http://www.susd12.org/content/sunnyside-district.

Sykut, Jamie. 2005. "Object Method Connections: Pestalozzi to Oswego (Sheldon)." http://www.oswego.edu.

The Telegraph. 2005. "Girl, 10, Used Geography Lesson to Save Lives." *The Telegraph.* http://www.telegraph.co.uk/news/1480192/Girl-10-used-geography-lesson-to-save-lives.html.

Trends in International Mathematics and Science Study. 2014. "Overview." Accessed July 16. http://nces.ed.gov/TIMSS/.

United States Department of Commerce. 2011. "STEM: Good Jobs Now and for the Future." http://www.esa.doc.gov/sites/default/files/reports/documents/stemfinalyjuly14_1.pdf.

United States Department of Education. 2004. "Highlights from the Trends in International Mathematics and Science Study: TIMMS 2003." Washington DC: National Center for Education Statistics.

Vacca, Richard T., and Jo Anne Vacca. 1999. *Content Area Reading: Literacy and Learning Across the Curriculum.* New York, NY: Addison-Wesley Educational Publishers, Inc.

Wist, Caroline C. 2014. "Putting it all Together: Understanding the Research Behind Interactive Notebooks." Accessed July 2.

Young Scientist Challenge. 2014. "About." Accessed July 2. http://www.youngscientistchallenge.com/about.

Recommended Literature

American Association for the Advancement of Science 2003. *Atlas of Science Literacy: Project 2061, Volumes 1 and 2*. Washington, DC: AAAS Press.

Chitman-Booker, Lakenna, and Kathleen Kopp. 2013. *The 5E's of Inquiry-based Science*. Huntington Beach, CA: Shell Education.

Donovan, M. Suzanne, and John D. Bransford. 2005. *How Students Learn: Science in the Classroom*. Washington, DC: The National Academies Press.

Fulton, Lori, and Brian Campbell. 2014. *Science Notebooks: Writing About Inquiry*, 2nd ed. Portsmouth, NH: Heinemann.

Hand, Brian, Lori Norton-Meier, Jay Staker, and Jody Blintz. 2009. *Negotiating Science: The Critical Role of Argument in Student Inquiry, Grades 5-10*. Portsmouth, NH: Heinemann.

Houghton Mifflin Harcourt. *ScienceSaurus*. Available at http://www.greatsource.com/.

Kopp, Kathleen. 2012. *Everyday Content-area Writing: Write-to-Learn Strategies for Grades 3-5*. Gainesville, FL: Maupin House Publishing, Inc.

Moomaw, Sally. 2013. *Teaching STEM in the Early Years: Activities for Integrating Science, Technology, Engineering, and Mathematics*. St. Paul, MN: Redleaf Press.

Westphal, Laurie. 2009. *Differentiating Instruction with Menus: Science*. Waco, TX: Prufrock Press, Inc.

Zembal-Saul, Carla, Katherine L. McNeill, and Kimber Hershberger. 2012. *What's Your Evidence?: Engaging K-5 Children in Constructing Explanations in Science*. New York, NY: Pearson.

Recommended URLs

Chapter 2

AIMS Education Foundation
http://www.aimsedu.org/

Arcadia National Park
http://www.nps.gov/acad/index.htm

College of the Atlantic
http://www.coa.edu/index.htm

Edmodo
https://www.edmodo.com/ *requires login*

Edutopia
http://www.edutopia.org/

Jason Project
http://www.jasonproject.org/

National Aeronautic and Space Administration (NASA)
http://www.nasa.gov

National Geographic for Teachers
http://education.nationalgeographic.com/education/?ar_a=1

National Oceanic and Atmospheric Association (NOAA)
http://www.oar.noaa.gov/k12

National Science Teachers Association
http://www.nsta.org/

Pinterest
https://pinterest.com/

Project WET
http://projectwet.org/

Science Foundation Arizona
http://www.sfaz.org/

Chapter 3

American Chemical Society, Centers for Disease Control and Prevention
http://www.acs.org/, http:www.cdc.gov

Electrical Safety Foundation International
http://www.esfi.org/

Chapter 4

Focus Magazine
http://sciencefocus.com/

Next Generation Science Standards
http://www.nextgenscience.org/ or http://www.nsta.org/about/
standardsupdate/

Scientific American
http://www.scientificamerican.com/

Chapter 5

American Association for the Advancement of Science, Project 2061
http://www.project2061.org/

Chapter 6

Common Core State Standards
http://www.corestandards.org/

Chapter 7

Cool Math
http://www.coolmath.com/graphit/

Discovery Education
http://www.discoveryeducation.com/

National Library of Virtual Manipulatives
http://nlvm.usu.edu/

PUMAS (Practical Uses of Math and Science)
https://pumas.gsfc.nasa.gov/

Chapter 8

Field Trip Factory
http://www.fieldtripfactory.com/

Intel Science Talent Search
http://www.intel.com/content/www/us/en/education/competions/science-talent-search.html

National Engineers Week Future City Competition
http://www.exploravision.org/

Odyssey of the Mind
http://www.odysseyofthemind.com/

Purdue University
http://www.purdue.edu/discoverypark/learningcenter/resources/outreach.php

Siemens Westinghouse Competition
http://www.siemens-foundation.org/en/

University of California, Los Angeles (UCLA)
http://www.k12outreach.ucla.edu/

University of Minnesota College of Science and Engineering
https://cse.umn.edu/k12/index.php

Chapter 9

eRubric Assistant
http://emarkingassistant.com/compare-emarking-assistant-and-erubric-assistant-essay-marking-software/erubric-assistant-free-rubric-generator/

Performance Assessment Links in Science (PALS)
http://pals.sri.com/

Chapter 10

RubiStar
http://rubistar.4teachers.org/

Science Competitions

Dupont Challenge
http://thechallenge.dupont.com/

ExploraVision
http://www.exploravision.org/

Future City Competition
http://futurecity.org/

Intel Science Talent Search
http://www.intel.com/content/www/us/en/education/competitions/science-talent-search.html

Kids Science Challenge
http://www.kidsciencechallenge.com/index.php

Odyssey of the Mind
http://www.odysseyofthemind.com/learn_more.php

Siemens Westinghouse Competition
http://www.siemens-foundation.org/en/competition.htm

Young Scientist Challenge
http://www.youngscientistchallenge.com/

Instructional Resources

This appendix contains several different resources for use in your classroom. While blank templates are not provided, these can be easily adapted to fit your class's instructional needs.

The first resource provided is the Sample Primary Checklist for Physical Science Concepts. Although specifically designed for physical science, this chart could be easily adapted for any science topic.

Next, you will find a Sample Rating Scale for an Animal Adaptation Story as well as a Sample Rubric for Unique Animal and Zoo Placard Summary. Both of these resources are intended to be used with a science-related story, but can be adapted for use in any content area.

Following these resources, you will find a Sample Summative Writing Project and Sample Summative Writing Project Evaluation. These two resources are on the topic of ecosystems, however the content can be easily modified to fit the needs of your classroom.

The last two resources you will find are a Sample Science Fair Timeline and Sample Science Fair Project Evaluation Scale. These samples are generic and ready-to-use and therefore do not require a great deal of modification.

Sample Primary Checklist for Physical Science Concepts

Student Name:	Yes	Somewhat	No
I can describe how objects move.			
I can show two different ways that objects move.			
I can show how to change an object's movement with a push.			
I can show how to change an object's movement with a pull.			
I can explain three ways that different objects move when I look at a picture of a park.			

Sample Rating Scale for Animal Adaptation Story

Student Name:		Date:	

Project Title: Animal Adaptation Story

Criteria	Below Expectation	Meets Expectation	Exceeds Expectation
Story is written from the perspective of the animal	0 1	2 3	4
Story includes a beginning, middle, and end	0 1	2 3	4
Grammar, spelling, and punctuation are correct	0 1	2 3	4
Identification of desert adaptations (3 or more) is clear, and naturally occurs in the story	0 1	2 3	4
Story includes one or more dangers to the animal as a desert-dweller	0 1	2 3	4
Story describes how the animal uses one or more adaptations to survive the dangers or challenges	0 1	2 3	4
Student clearly demonstrates understanding of desert-dwelling animal adaptations	0 1	2 3	4
Overall project evaluation (on topic, complete, and interesting to read)	0 1	2 3	4
Score			_____ / 32
Student Comments:			
Teacher Comments:			

Sample Rubric for Unique Animal and Zoo Placard Summary

Criteria	Level 4	Level 3	Level 2	Level 1	Level 0
Content: Illustration or 3D Animal Model	Accurate, detailed illustration or 3D animal model	Complete illustration or 3D animal model, lacks clear details	Incomplete or inaccurate illustration or 3D animal model with few details	Incomplete and inaccurate illustration or 3D animal model with few, if any, details	No illustration or 3D animal model
Content: Placard Summary	Complete and accurate details (three or more) to describe how animal is adapted to live in a desert	Complete description of how animal is adapted to live in a desert with two details	Incomplete description of how animal is adapted to live in a desert; only one detail	Incomplete description of how animal is adapted to live in a desert; off topic	No description of how animal is adapted to live in a desert
	Clear, concise, appropriate use of content-specific vocabulary	Appropriate use of content-specific vocabulary	Some appropriate use of content-specific vocabulary	Little appropriate use of content-specific vocabulary	No appropriate use of content-specific vocabulary
Organization	Extremely well organized with a clear beginning, middle, and end	Well organized with a beginning, middle, or end, but not all three	Some attempt at organization, but lacks a clear beginning, middle, and end	Disorganized, with no clear beginning, middle, or end	No organizational structure evident
Conventions	Few, if any, errors in spelling, grammar, and/or punctuation	Some errors in spelling, grammar, and/or punctuation that do not detract from the overall summary	Several errors in spelling, grammar, and/or punctuation that may detract from the overall summary	Several errors in spelling, grammar, and/or punctuation that detract from the overall summary	No attempt made to use correct grammar, spelling, and/or punctuation

Sample Summative Writing Project

News Alert: Ecosystem!

Situation: You are a reporter for a local area news channel. Your manager has asked you and a team of reporters to do a special story on an ecosystem that is being threatened by an invasive species.

Task: Write a news report about a real or fictional ecosystem. Explain how an invasive species is threatening this ecosystem.

Required: Your ecosystem news report must:

- Have a clear beginning, middle, and end
- Mention an invasive organism (plant or animal), or pollution
- Include a description as to how the invasive organism or pollution negatively affects the ecosystem
- Name native plants and animals and identify them as producers, consumers, or decomposers; and herbivores, carnivores, or omnivores
- Include an energy source

Directions

Prewriting

1. Decide on an ecosystem. It may be real or fictional.

2. List the organisms (plants and animals) you will include in your ecosystem. Identify the invasive organism. Identify the energy source.

3. Plan your news report.

Drafting

4. Write your news report. Include mention of all the necessary parts of the ecosystem (see above).

5. Develop the story. This news report is really about the invasive organism.

6. Wrap up your story. What will scientists do next, or what is their plan to save the ecosystem?

Revise and Edit

7. Be sure your story has a clear beginning, middle, and end.

8. Revise some verbs to show more specific action.

Publish your story.

9. Read your news report. Everyone in the group should have a part.

Sample Summative Writing Project Evaluation

Directions: Review these criteria. This is how your ecosystem news report will be graded. Review your news report. Be sure it has all the makings of an exciting news story!

Criteria	Possible	Earned
1. The project demonstrated a thorough understanding of an ecosystem.	10	_____
2. The project was neat, well organized, and an example of the student's best work.	20	_____
3. The ecosystem included an invasive organism (plant OR animal) OR pollution.	10	_____
4. It was apparent OR explained how the invasive organism/pollution negatively affected the ecosystem.	10	_____
5. The ecosystem contained native plants.	10	_____
6. The ecosystem contained native animals.	10	_____
7. The ecosystem contained a producer, consumer, and decomposer (5 pts each).	15	_____
8. The energy source for the ecosystem was apparent or explained when presented.	5	_____
9. Students clearly identified the animals in their ecosystem as carnivores, herbivores, or omnivores.	10	_____
TOTAL:	100	_____

Teacher Comments:

_____

_____

_____